METRIC MEASU

with sugar cubes, paper, straws, string, and simple things

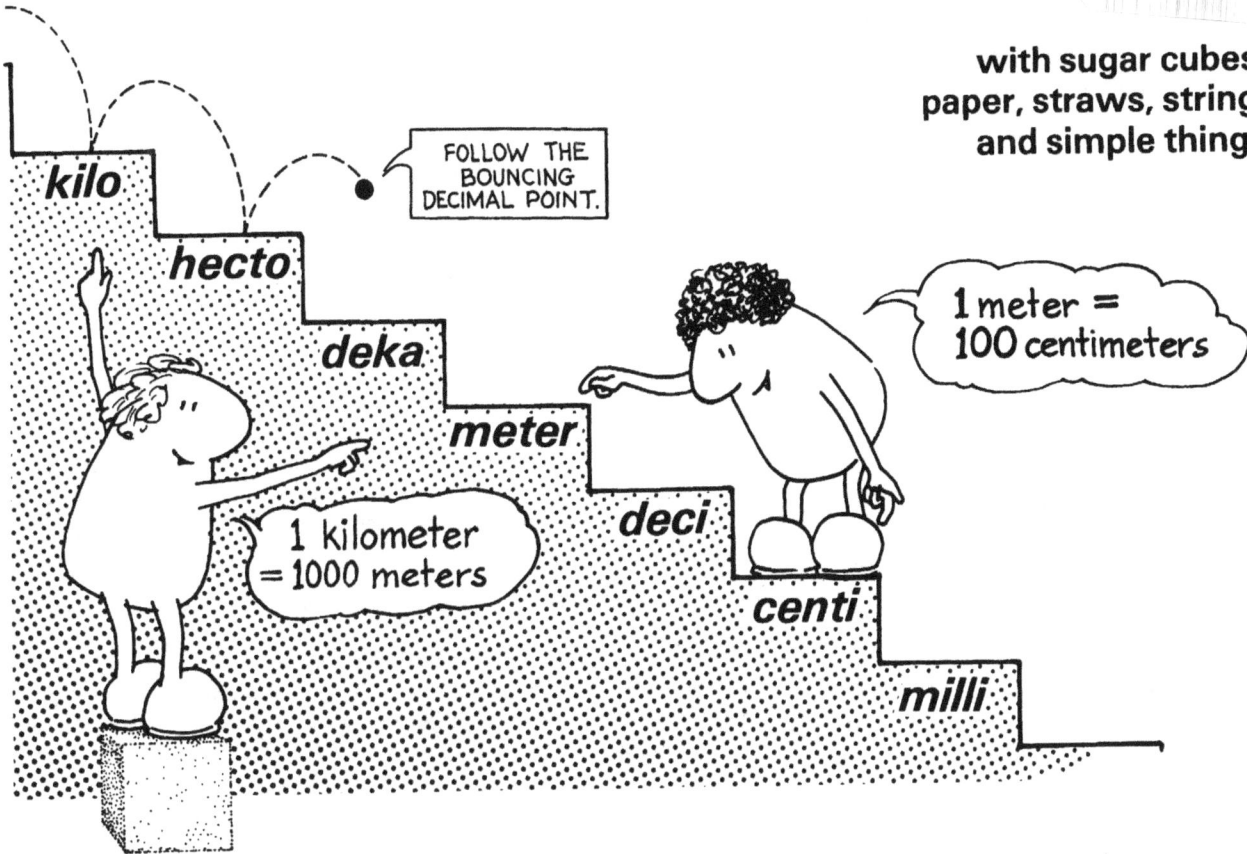

FOLLOW THE BOUNCING DECIMAL POINT.

kilo
hecto
deka
meter
deci
centi
milli

1 meter = 100 centimeters

1 kilometer = 1000 meters

SCIENCE WITH SIMPLE THINGS SERIES

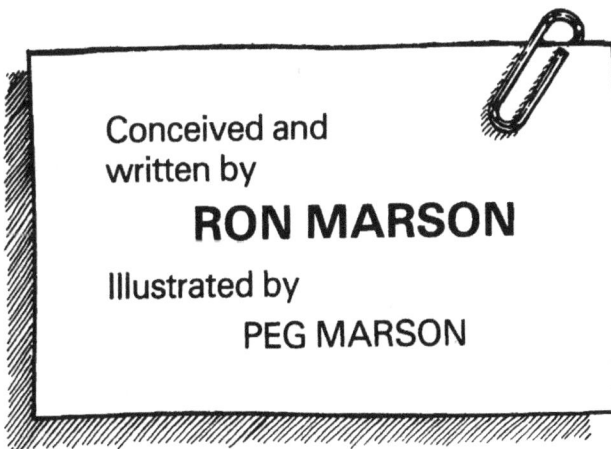

Conceived and written by
RON MARSON

Illustrated by
PEG MARSON

342 S Plumas Street
Willows, CA 95988

TOPS LEARNING SYSTEMS

www.topscience.org

WHAT CAN YOU COPY?

Dear Educator,

Please honor our copyright restrictions. We offer liberal options and guidelines below with the intention of balancing your needs with ours. When you buy these labs and use them for your own teaching, you sustain our work. If you "loan" or circulate copies to others without compensating TOPS, you squeeze us financially, and make it harder for our small non-profit to survive. Our well-being rests in your hands. Please help us keep our low-cost, creative lessons available to students everywhere. Thank you!

PURCHASE, ROYALTY and LICENSE OPTIONS

TEACHERS, HOMESCHOOLERS, LIBRARIES:

We do all we can to keep our prices low. Like any business, we have ongoing expenses to meet. We trust our users to observe the terms of our copyright restrictions. While we prefer that all users purchase their own TOPS labs, we accept that real-life situations sometimes call for flexibility.

Reselling, trading, or loaning our materials is prohibited unless one or both parties contribute an Honor System Royalty as fair compensation for value received. We suggest the following amounts – let your conscience be your guide.

HONOR SYSTEM ROYALTIES: If making copies from a library, or sharing copies with colleagues, please calculate their value at 50 cents per lesson, or 25 cents for homeschoolers. This contribution may be made at our website or by mail (addresses at the bottom of this page). Any additional tax-deductible contributions to make our ongoing work possible will be accepted gratefully and used well.

Please follow through promptly on your good intentions. Stay legal, and do the right thing.

SCHOOLS, DISTRICTS, and HOMESCHOOL CO-OPS:

PURCHASE Option: Order a book in quantities equal to the number of target classrooms or homes, and receive quantity discounts. If you order 5 books or downloads, for example, then you have unrestricted use of this curriculum for any 5 classrooms or families per year for the life of your institution or co-op.

2-9 copies of any title: 90% of current catalog price + shipping.

10+ copies of any title: 80% of current catalog price + shipping.

ROYALTY/LICENSE Option: Purchase just one book or download *plus* photocopy or printing rights for a designated number of classrooms or families. If you pay for 5 additional Licenses, for example, then you have purchased reproduction rights for an entire book or download edition for any **6** classrooms or families per year for the life of your institution or co-op.

1-9 Licenses: 70% of current catalog price per designated classroom or home.

10+ Licenses: 60% of current catalog price per designated classroom or home.

WORKSHOPS and TEACHER TRAINING PROGRAMS:

We are grateful to all of you who spread the word about TOPS. Please limit copies to only those lessons you will be using, and collect all copyrighted materials afterward. No take-home copies, please. Copies of copies are strictly prohibited.

For licensing, honor system royalty payments, contact: **www.TOPScience.org**; or **TOPS Learning Systems 342 S Plumas St, Willows CA 95988**; or inquire at **customerservice@topscience.org**

ISBN 978 - 0 - 941008 - 35 - 8

CONTENTS

GETTING IT TOGETHER

You hold within your hands a **complete teaching resource.** This book contains 20 reproducible hands-on science lessons together with all necessary information to help you teach each lesson successfully. All you add are the simple materials listed at the bottom of the page.

Look it over. This modest list contains everything you need to teach **every** lesson. Most of the materials you already have. Get the rest from your local supermarket or have your students bring the required items from home.

Each item is **listed in order** of first appearance in the student activities. To start getting it together, begin at the top of this list and work down. Gather everything at once, or collect materials as your students progress through each lesson.

Needed quantities depend on several factors: how you teach, how many students you have and how you organize them into activity groups. The numbers listed by each item correspond to the main teaching strategies in use today. Find the one that suits your teaching style and gather quantities accordingly.

From time to time the teaching notes contain suggestions for additional activities called EXTENSIONS. Materials for these optional experiments are not listed here nor under MATERIALS in the teaching notes. Read instead the extension itself to find out what new materials, if any, are required.

Once you collect the needed materials, place them on an equipment table or on open shelves that are accessible to your students. Items of special value may require a locked cabinet or a special check-out box near the teacher's desk.

Many of the materials you use in this module are used in other TOPS modules as well. As you continue with other TOPS modules and build your inventory, you'll find that gathering materials requires less and less effort!

Q₁

Resource Center
Activity Corner
Parent-Child Activity
Demonstrations

There is enough material so that 1 student or group of students can complete all the activities.

If you multiply Q₁ by 2, then there will be enough materials for two groups to work on the same activity or, perhaps, for three or more groups to simultaneously work on different activities.

Q₂

Individualized Approach

Initial activities require almost as much duplication as the traditional approach. But quantities soon drop off as groups "spread out" within the module, doing different activities at different times.

Students group naturally and informally according to academic or social preferences. Group membership tends to change as slower members fall back into slower groups and faster members move up into faster groups.

Quantities in Q₂ assume a total class size of about 30 students working in 10 groups of 3 each. Modify as necessary to fit your own particular requirements.

Q₃

Traditional Class Lessons

The teacher introduces each lesson to the class as a whole, then everyone does the activity together. Time at the end of the period is reserved for summarizing and reinforcing key concepts.

Quantities in Q₃ again assume a class size of about 30 students working in groups of 3. The numbers are sometimes higher than Q₂ because greater duplication of materials is needed when everyone works simultaneously on the same worksheet.

MATERIALS

Q₁	/Q₂	/Q₃	
10	/70	/70	sheets of lined notebook paper— square corners preferred
1	/9	/9	pairs of scissors
30	/270	/270	sugar cubes equivalent to 1 teaspoon — see teaching notes 1
1	/9	/9	rolls of cellophane tape
1 box			paper clips
1 roll	/ 2 rolls		adding machine tape
1 ball			kite string
1	/3	/9	hand calculators (optional)
3	/27	/27	3x5 index cards
1	/3	/9	large grocery bags
4	/36	/36	plastic soda straws
5	/40	/40	straight pins — see teaching notes 7
1	/30	/30	wooden spring-action clothespins
2	/35	/40	soda pop cans with pull tabs attached

Q₁	/Q₂	/Q₃	
10	/130	/150	small styrofoam cups
10	/100	/250	U.S. pennies minted after 1982
2 ea.	/9 ea.	/18 ea.	pre-1982 U.S. pennies plus U.S. nickels and quarters with any date
3	/12	/24	U.S. dimes
1 handful			uncooked long-grain rice
1	/3	/9	deep plastic tubs — dishwashing size
10	/200	/200	sheets of medium to heavy 8½x11 paper — see teaching notes 13
1 pkg.			table salt
1 pkg.			plastic sandwich bags
1	/3	/9	empty quart milk cartons
1 pkg.			granulated sugar
1	/6	/18	spoons
1 pkg.			corn meal

SEQUENCING ACTIVITIES

This logic tree shows how all the worksheets in this module tie together. In general, students begin at the trunk of the tree and work up through the related branches. As the diagram suggests, the way to upper level activities leads up from lower level activities.

At the teacher's discretion, certain activities can be omitted or sequences changed to meet specific class needs. The only activities that *must* be completed in sequence are indicated by leaves that are linked vertically with an *open space* in between. In this case the lower activity is a prerequisite to the upper.

When possible, students should complete the worksheets in numerical sequence, from 1 to 20. If time is short, however, or certain students need to catch up, you can use the logic tree to identify concept-related *horizontal* activities. Some of these might be omitted since they serve only to reinforce learned concepts rather than to introduce new ones.

On the other hand, if students complete all the activities at a certain horizontal concept level, then experience difficulty at the next higher level, you might go back down the logic tree to have students repeat specific key activities for greater reinforcement.

For whatever reason, when you wish to make sequence changes, you'll find this logic tree a valuable reference. Parentheses in the upper right corner of each student worksheet allow you this flexibility: they are left blank so you can pencil in sequence numbers of your own choosing.

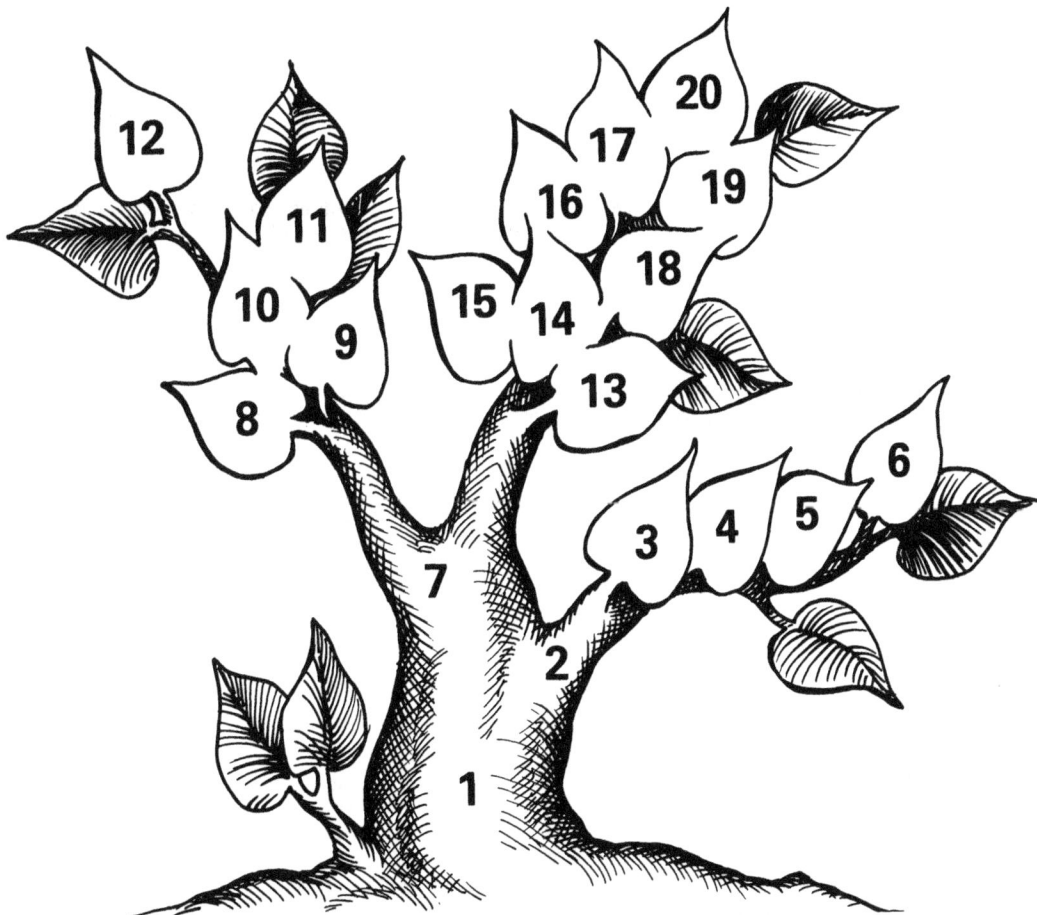

METRIC MEASURING 35

BUILDING AN EFFECTIVE TEACHING STRATEGY

No teaching strategy is totally effective in all classrooms situations. This module is flexibly arranged to adapt to a wide *range* of teaching possibilities. Design your own strategy: select options listed below that best fit your own needs and meet the needs of your students.

A. Classroom Organization

1 RESOURCE CENTER
Worksheets and science materials are placed in a special resource area. Students come from the classroom to work independently on science activities. Teachers or aides are available to assist students as the need arises.

2 ACTIVITY CORNER
This operates like the resource center, except a special area is designated *within* the classroom itself. Students come here to do science experiments after their regular class work is completed.

3 PARENT-CHILD ACTIVITY
A parent, teacher or aide works with one or more students in a tutorial relationship. This may occur during or after school hours or in the home.

4 TEACHER DEMONSTRATION
The teacher performs experiments in front of the class, inviting occasional student participation. This approach is often used for younger children who do not have sufficient manual dexterity to manipulate the materials.

5 INDIVIDUALIZED ACTIVITY
Students proceed through the worksheets at their own pace. Those working on the same activity informally group together, reducing substantially the total number of experiments going on in the classroom. The teacher acts as a learning supervisor, responding to questions and problems as they arise within the context of class activity. After the most advanced students complete all the worksheets, the class moves on to a new module *together*.

6 TRADITIONAL CLASS TOGETHERNESS
Each activity constitutes a specific lesson to be completed during a specified time frame. The teacher introduces the activity to all the students together, then breaks the class into managable lab groups that each do the same experiment. A class discussion sometimes follows to summarize key concepts and provide lesson closure.

B. Reproduction of Activity Sheets

1 OVERHEAD PROJECTION
Place each worksheet directly on an opaque projector or prepare a transparency.

2 ACTIVITY CARDS
Make 2 or 3 photocopies of each worksheet. Plasticize them to make durable full page activity cards. File these in an activity folder or display them on a bulletin board or wall.

3 WORKSHEETS
Duplicate enough copies to provide each student with a worksheet. These can be photocopied directly or thermofaxed onto a master ditto, then run off on a spirit duplicator. Distribute copies to students directly. Or place each set in a separate folder and file them in a box so students can use them as needed.

C. Evaluation

1 PASS/NO-PASS CHECKPOINTS
Daily write-ups are evaluated by the student and teacher together *in class*. If the student demonstrates reasonable effort commensurate with ability, the write-up is simply checked off, either in a grade book or on a progress chart attached to each student's personal assignment folder kept on file in class.

2 GRADED ASSIGNMENTS
Write-ups are handed in by each student as completed, graded by a teacher or an aide and then returned to the student.

3 QUIZES
The teacher gives a quiz (written or oral) after each activity. Questions for the quiz are taken from the "Evaluation" sections in the teaching notes.

4 INFORMAL OBSERVATION
The teacher takes mental note of active participators who work to capacity and of inactive onlookers who waste time. Grades are awarded accordingly.

5 EXAMS
An exam given to all students at the same time covers key concepts from activities that all students have completed and reviewed. Questions come from the "Evaluation" sections in the teaching notes.

Among the teaching options listed above, we recommend the combination A5 - B3 - C1 - C4 - C5. This approach combines elements from two opposite teaching strategies in a most effective way: it allows for individual differences while maintaining traditional class togetherness.

Would such an approach work in your own classroom? The "Dairy of a Teacher" which follows may help you visualize the answer. It is based on my own classroom experience.

DIARY OF A TEACHER

THE DAY BEFORE

Tomorrow is the first day of school. With anxiety and anticipation you check to see that everything in your classroom is in good order.

You have already duplicated 30 copies each of the first few lessons in this TOPS module. They lay on your desk, each in a manilla folder marked with large numbers 1, 2 and 3. You make a mental note to bring a large box and a brick to school tomorrow. These will prop the folders upright like a vertical file, and help keep you organized as you add additional folders.

You wonder how you ever managed without manilla folders. Already you have printed each student's name on a fresh new assignment folder and stapled a sheet of graph paper to the inside cover to track each student's progress. When they arrive tomorrow, you will surprise them with a worksheet, a file folder and the simple instruction, "Get busy". You've already laid out the necessary materials on a table in the back of your room. You smile inside yourself; you haven't felt quite this prepared in years.

DAY 10

Your class has been humming now for 10 straight days. Perhaps not humming: buzzing more aptly describes the state of orderly confusion. Students have questions and problems to be sure. (You wish they would at least *read* the instructions before running to you for an explanation.) Still, the worksheets provide a firm sense of direction. Students know where they are and understand where they need to go.

Now that students understand your system, they come to class and get straight to work on what they were doing the day before. Just before lunch they tend to quit early, but at other times you have to pry them away from their experiments. You tell the slower ones to assign themselves homework to catch up and it seems to be working!

The assignment folders work well too. Students point with pride at the growing list of check-points you have marked off on their graph paper progress charts. As their folders expand so does their self confidence.

DAY 15

Today 2 groups of students who seem to be racing each other have completed all 20 activities. The bulk of your class remains 3 to 6 activities behind, with a few stragglers plus the new kid bringing up the rear.

You announce that individualized worksheet activity will end in 2 days. The most advanced students seem eager to work on several experiments of their own. You can follow that up with an "Extension" activity if time allows. You help the slow ones catch up by assigning three key-concept activities while skipping the rest.

There is a frenzy of activity as students rush to meet your deadline. They know that part of their grade is determined by the total number of activities they complete.

DAY 18

Today you kick back and relax. You have assigned several students to give reports on their original investigations. The rest of the period will be taken up with a film. For tomorrow you've planned a blackboard review of major module concepts. Then on Friday you'll finish off with an exam.

The kids are already bugging you about grades and asking what they will be studying next. You decide to give them a 3-part grade weighted equally on pace (number of lessons completed), attitude, and the exam. As to what they'll study next, you can't decide. Perhaps you'll let *them* decide.

Its already the fourth week into the school year and you don't even feel the strain. Activity-centered teaching seems so natural and easy. You respond to questions that kids have instead of the other way around. You ask yourself why you never taught this way before.

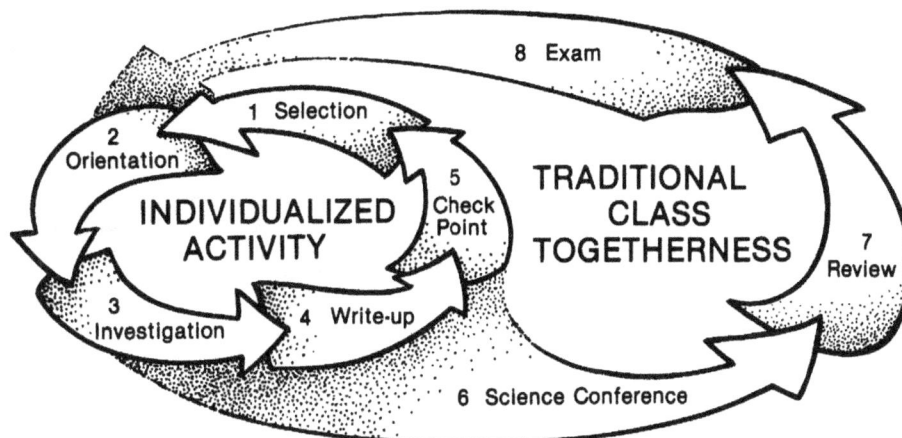

1. **SELECTION.** Students generally select worksheets in the order you specify. They should be allowed to skip a task that is not challenging, however, or repeat a task with doubtful results. When possible, encourage students to do original investigations that go beyond or replace particular activites.

2. **ORIENTATION.** Good students will simply read worksheet instructions and understand what to do. Others will require further verbal interpretation. Identify poor reader in your class. When they ask, "What does this mean?" they may be asking in reality, "Will you please read these instructions aloud?"

3. **INVESTIGATION.** Students observe, hypothesize, predict, test and analyze, often following their own experimental strategies. The teacher provides assistance where needed and the students help each other. When necessary, the teacher may interrupt individual activity to discuss problems or concepts of general class interest.

4. **WRITE-UP.** Worksheets ask students to explain the how and why of things. Answers should be brief and to the point. Students may accelerate their pace by completing these out of class.

5. **CHECK POINT.** The student and teacher evaluate each write-up together on a pass/no-pass basis. If the student has made reasonable effort consistent with individual ability, the write-up is checked off on a progress chart and included in the student's personal assignment folder kept on file in class.

6. **SCIENCE CONFERENCE.** After individualized activity has ended, students come together to discuss experiments of general interest. Those who did original investigations give brief reports. Slower students learn about the later activities completed only by faster students. Newspaper articles are read that relate to the topic of study. The conference is open to speech making, debate, films, celebration, whatever.

7. **REVIEW.** Important concepts are discussed and applied to problem solving in preparation for the module exam.

8. **EXAM.** Evaluation questions are written in the teaching notes that accompany each activity. They determine if students understand key concepts developed in the worksheets. Students who finish the test early begin work on the first activity in the next new module.

D

LONG-RANGE OBJECTIVES

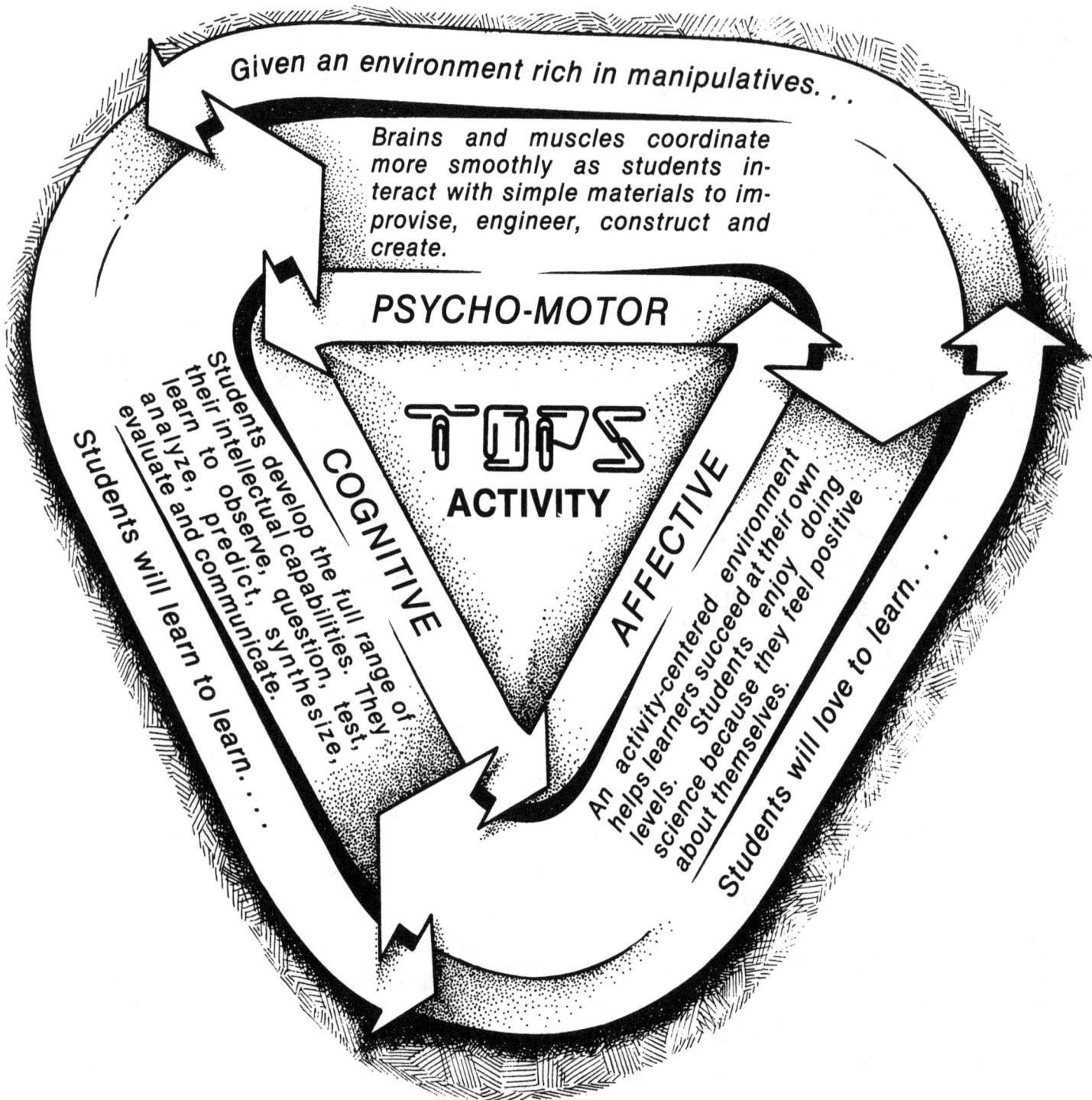

Given an environment rich in manipulatives. . .

Brains and muscles coordinate more smoothly as students interact with simple materials to improvise, engineer, construct and create.

PSYCHO-MOTOR

TOPS
ACTIVITY

COGNITIVE

Students develop the full range of their intellectual capabilities. They learn to observe, question, test, predict, synthesize, analyze, evaluate and communicate.

Students will learn to learn. . . .

AFFECTIVE

An activity-centered environment helps learners succeed at their own levels. Students enjoy doing science because they feel positive about themselves.

Students will love to learn. . . .

GAINING A WHOLE PERSPECTIVE

Science is an interconnected fabric of ideas woven into broad and harmonious patterns. Use "Extensions" in the teaching notes plus the outline presented below to help your students grasp the big ideas—to appreciate the fabric of science as a unified whole.

Resolved: that the US Congress should pass a law requiring that all commercial products and road signs be written in metric measure ONLY.
Do you agree or disagree? Hold a **class debate**.

Related TOPS modules that provide additional hands-on measuring experience using simple materials include:

02 Measuring Length
03 Graphing
06 Metric Measure
36 More Metrics

There are **metric prefixes** much larger than KILO and much smaller than MILLI. Extend this list as far as you can in both directions:

kilo $= 10^3$ or 1000
hecto $= 10^2$ or 100
deka $= 10^1$ or 10
$\quad\quad\quad 10^0$ or 1
deci $= 10^{-1}$ or .1
centi $= 10^{-2}$ or .01
milli $= 10^{-3}$ or .001

METRIC MEASURING 35

How did people measure distance before rulers were invented? Study the origin of terms like cubit, span, fathom and inch.

As a measuring expert, your assignment is to publish a GLOSSARY OF VISUAL METRIC IMAGES. A measuring beginner should be able to look up any common metric unit in your **glossary** and find it expressed in **familiar terms**.

Ex: meter — as wide as a doorway; about 1 giant step.

How cold is water when it freezes? How hot when it boils? Answer in **three units of temperature:** Farenheit, Celsius and Kelvin. (Which units do scientists prefer?)

Read **FLATLAND** by Edwin A. Abbott. Explore the fascinating 2-dimensional world of the flatlanders from your own 3-dimensional perspective. Use this experience to stretch your mind into 4 dimensions!

TEACHING NOTES
For Activities 1-20

Why has almost the whole world, even the English adopted the metric system? That's not difficult to understand. Converting from one metric unit into another is as easy as moving a decimal point. (They hardly weigh anything at all!)

Why is only the United States still clinging to 12 inches make a foot, 3 feet make a yard, 1,760 yards make a mile . . . ? That's hard to fathom. In or out of the U.S., almost anyone who understands metrics will prefer using metrics.

To begin to understand, let's make sure your class appreciates the computing power of decimals. Turn to the DECIMAL FLOW CHART between Teaching Notes 3 and 4.

1. Kilo is another way of saying 1,000, hecto means 100 and so on, right down the stairs. Notice that the "ones" step has no prefix at all. Here you can substitute any kind of measure you want—dollars, meters, feet, apples, elephants—anything. Thus, a *deka*elephant means *10* elephants, probably more than you'd want to feed; a *deci*foot means *1/10* foot, a little longer than an inch.

3. You translate prefixes into numbers again, this time within the context of dollars. The cube is not really needed here, but we ask students to throw it anyway. This helps them learn how to write equations from the top of the cube.

5. Your students are going to need lots of help to make it through this step. For the first time they must convert from one set of units (dollars) into another set of units (whatever turns up on their cube). So, call a temporary halt to class activity and hold this class discussion:

Metric stairs help you change one unit of measure into another without losing or gaining anything. To maintain this equality, just follow the arrows and move the decimal point accordingly.

To go *down* the stairs and stay equal, move the decimal to the *right* (multiply) in the same direction you descend. Thus,

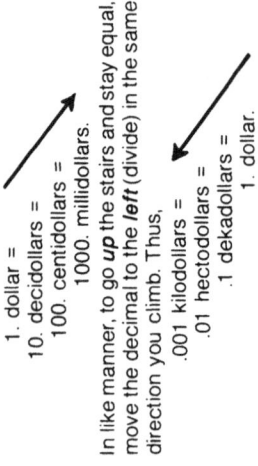

1. dollar =
10. decidollars =
100. centidollars =
1000. millidollars.

In like manner, to go *up* the stairs and stay equal, move the decimal to the *left* (divide) in the same direction you climb. Thus,

.001 kilodollars =
.01 hectodollars =
.1 dekadollars =
1. dollar.

Use as many blackboard examples as necessary to reinforce this division-multiplication process.

6. By tossing *both* cubes your students will learn they can use the decimal stairs to find equivalents from *any* step (not just the middle one) to any other step.

The process is always the same: begin with what you're given, then multiply down or divide up, moving the decimal right or left until you reach the unit you want. As your students repeat this process over and over, they will soon forget about the staircase analogy. Of course there are 100 centimeters in a meter. They'll say it's obvious!

7. This a cooperative, self-checking metric conversion game. As your students play it, metric logic will become an integral part of their thinking.

Worksheet Answers

1.
kilo = 1000		deci = .1
hecto = 100		centi = .01
deka = 10		milli = .001

3.
1 kilodollar = $1,000	1 decidollar = 10¢
1 hectodollar = $100	1 centidollar = 1¢
1 dekadollar = $10	1 millidollar = 1/10¢

5.
$1 = .001 kilodollars	$1 = 10 decidollars
$1 = .01 hectodollars	$1 = 100 centidollars
$1 = .1 dekadollars	$1 = 1000 millidollars

6. There are 36 possible combinations that can turn up on the cubes. Here are just a few.

   1 kilodollar = 1,000,000 millidollars
   1 hectodollar = 10,000 centidollars
   1 millidollar = .000001 kilodollars
   1 centidollar = .0001 hectodollars
   1 dekadollar = 1 dekadollar

Evaluation

Q: Draw a metric staircase with 7 steps. Label all numbers and prefixes.

A: Students should draw and label a staircase similar to the one on their worksheet.

Materials

☐ Lined notebook paper.
☐ Scissors.
☐ Sugar cubes. You must use cube shapes (not bricks) equivalent to 1 teaspoon each (not ½ teaspoon). C and H brand "cubelets" are available in the West; other brands in the East. Match yours against this actual-size pattern. If you can't locate teaspoon cubes in your area, order direct from TOPS. We'll ship 252 cubes per 2 lb box (specify quantity), and bill you our cost plus shipping and handling.
☐ Cellophane tape.
☐ A paper clip.

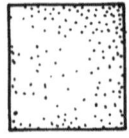

Task Objective (TO) understand the language of metric prefixes. To learn how to make metric conversions by moving the decimal point.

NAME: _____ CLASS: _____

Metric Measuring( )₁

METRIC STAIRS (1)

DIVIDE ← → MULTIPLY

| kilo 1000 | hecto 100 | deka 10 | (dollar) 1 | deci .1 | centi .01 | milli .001 |

000001.000000

MEMORIZE THIS AND YOU TOO CAN BE A METRIC WIZ!

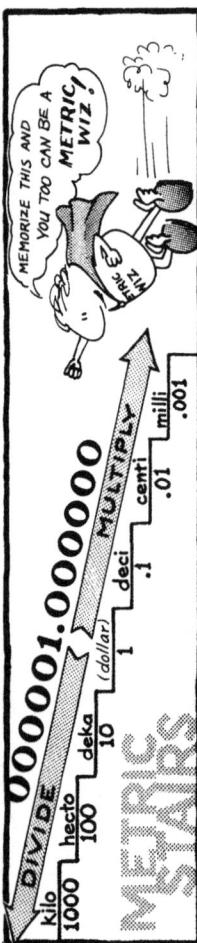

1 On a piece of lined notebook paper, write 6 equations, 1 for each metric step. WRITE THIS TOP STEP FIRST.
kilo = 1000

2 Cut out this cube pattern. Fold and tape it around a SUGAR CUBE! PUT A SUGAR CUBE INSIDE.
(Cube faces: 1 MILLI dollar, 1 CENTI dollar, 1 KILO dollar, 1 DECI dollar, 1 HECTO dollar, 1 DEKA dollar)

3 Roll 6 *different* cash amounts. Write how much each is, just like the example.
EXAMPLE: 1 dekadollar = 10 dollars

4 Cut out this cube pattern with boxes. Fold and tape it around a sugar cube like you did with the first.
(Cube faces: DECI dollars, DEKA dollars, HECTO dollars, KILO dollars, CENTI dollars, MILLI dollars) (BOXES)

5 Throw this "box" cube 8 times. Each time it lands find the missing number that *makes one dollar*. (STUDY THE EXAMPLE!)
EXAMPLE: 1 dollar = 100 centidollars
1 dollar =
1 dollar =
1 dollar =

6 Toss *both* cubes 12 times. Copy each equation (first the "1" cube, then the "box" cube), then find the missing number.
STUDY THIS EXAMPLE
1 hectodollar = ☐☐ kilodollars

7 Two can play metric dice: roll one pair of dice; write down answers; compare. If answers agree, move a paper clip forward 1 space. If answers disagree, move back 2 spaces. AS A TEAM, CAN YOU MOVE THE PAPER CLIP FROM 0 to 10?
START 1 2 3 4 5 6 7 8 9 10 FINISH

TOPS LEARNING SYSTEMS

Teaching Notes 2

Sugar cubes are cheap, durable and have uniform size. They make ideal building blocks for fixing concepts of length, area, volume and mass firmly in mind.

Sugar cubes aren't cut to metric dimensions, of course. They don't need to be. Beginning with this activity and continuing through activity 9, we logically develop a "sweetmeasure" system that mimics the official metric system, sugar cube for centimeter. Just as 100 centimeters make one meter, in this activity we say 100 sugar cubes laid end to end define "1 Length". In lesson 10, the transition from Lengths to meters will be smooth and easy. Simply put away your sugar cubes and think about metric water cubes instead.

1. The full-sized sweetruler in worksheet 2 (not its reduction to the left) measures sugar cubes here, straight lines in worksheet 3, rectangles in worksheet 4 and blocks in worksheet 5. This ruler will not measure these objects in exact whole numbers, as we've designed it to do, unless you reproduce it free of distortion. Unfortunately, some photocopiers lengthen or shorten images in one dimension without a proportional size change in the other dimension. (Believe it or not!)

Check to see if you can make true reproductions so that your copied sweetruler measures the copied lines, rectangles and blocks in the same exact whole numbers that are listed on the answer keys. If the distortion is unacceptable, try using the substitute sweetruler between worksheets 14 and 15. Because of its horizontal positon, it can perhaps be copied with greater accuracy. (And you'll reproduce graph paper for later use as well.)

If you find that unacceptable distortions still remain, alter our worksheet originals to compensate. Remember that your doctored originals will only work with your particular copy system. Other copy systems will likely introduce different distortion patterns.

2. If work space is cramped, so that tapes spill over into adjacent work area, ask your class to treat them like scrolls. Provide paper clips to keep the ends from unrolling.

3. Draw each sugar cube calibration down to the bottom edge of the paper strip. Some may draw marks that wander above this edge.

If some sweetrulers are not well calibrated, don't be too concerned. We never require your students to measure accurately beyond the first 10 sugar cubes. We preprinted the first part of each ruler for this reason. Your whole class can use this same reliable mini-standard, even though the longer tapes may not be uniform when stretched to full length.

4. Stress the arbitrary nature of this unit. Instead of calling 100 sugar cubes "1 Length" we could have called this distance by some other name. (We almost called it a "sweetmeter", but subunits like "centisweetmeters" proved too clumsy.)

Want to ham it up? Christen the ruler as "1 Length" in a class ceremony filled with great pomp and circumstance. Your class will not likely forget the name or its arbitrary origins.

5-10. These questions place sweetmeasure within a metric context. Make sure your students do some hard thinking before you agree to help them. Don't allow them to get the thinking lazies, running to you every time an answer is not immediately apparent. Once formed, this habit is difficult to break. Instead, allow your students the satisfaction of figuring things out on their own—using the metric stairs in activity 1.

Evaluation

Q: a. Get a paper clip. Along the top edge of your test paper, make a ruler that measures in paper clips.
b. Use your ruler to measure the length of your desk in paper clips.

A: a. Students should mark off equal paper clip intervals along the top edge of their test papers like this:

b. If your desks have uniform length, answers should not vary by more than a paper clip or two. Accept no answer without its unit given in paper clips.

Materials
- □ Scissors.
- □ Strips of paper. Precut adding machine tape into lengths that are roughly 2 meters (6 feet) long. You can use your own height or arm span to make a rough estimate. Your students will trim off the excess. If adding machine tape is not available, try substituting a roll of smooth continuous paper towel that you can write on. You or your students will have to cut off long narrow strips from this roll, about 4 sugar cubes wide.
- □ Cellophane Tape.
- □ Sugar cubes.
- □ String.
- □ Activity sheet 1. Students may need to use the metric stairs as a reference.

(TO) construct a ruler that measures in sugar cubes. To invent a unit of measure that corresponds to meters in the metric system.

NAME:

CLASS:

Metric Measuring ( )2

SWEET LENGTHS

1 Cut out this ruler.

2 Tape this ruler to the *lower left* corner of a long strip of paper.
PLACE IT ON THE EDGE.

How many sugar lumps fit between 0 and 10 on this ruler?
10 lumps

3 Use a row of 10 sugar lumps to number a full "sweetruler" with 100 lumps.
MAKE A PENCIL MARK BETWEEN EACH LUMP AT THE EDGE OF THE PAPER.

4 Cut your sweetruler off at the 100 mark. Write "1 LENGTH" on your ruler in large bold letters.
I HEREBY NAME THEE ONE LENGTH
1 LENGTH

5 Your ruler is 1 Length. Is your room longer than a deka Length?
10 of these = 1 deciLength
most rooms are shorter

6 Your ruler is 1 Length: Name the distance equal to 100 sweetrulers laid end to end.
a hectoLength

7 Your ruler is 1 Length: Name the distance equal to 1000 sweetrulers laid end to end.
a kiloLength

8 Your ruler is 1 Length: Name the distance equal to 0.1 sweetrulers.
1 deciLength
How many sugar lumps measure this long?
10 sugar lumps

9 Your ruler is 1 Length: Name the distance equal to 0.01 sweetrulers.
1 centiLength
How many sugar lumps measure this long?
1 sugar lump

10 Your ruler is 1 Length: Name the distance equal to 0.001 long:
1 milliLength
How many milliLengths are in your sweetruler?
1,000 milliLengths

WRITE YOUR NAME ON YOUR SWEETRULER AND SAVE IT!

100 centiLengths = 1 Length
1 centiLength
0 1 2 3 4 5 6 7 8 9 10
SWEETRULER →

count units to find the correct measure.

— *This is 40 cL,
not 60 cL.* —

c. Report all measurements as decimals, not as mixed fractions.

1a-h. The measurements in the answer key to the left are *not* given in significant figures. If they were, you would have to express each measure to 3 digits—2 certain figures plus one estimated figure. The first line, for example would measure 3.00 cL, implying measuring certainty until you reach the estimated hundredths place.

There are answer boxes on either side of the lines in this worksheet. So conversion from one unit of measure into another will be necessary. There are . . .

10 milliLengths in 1 centiLenght,
10 deciLengths in 1 Length,
10 hectoLengths in 1 kiloLength

How do you know? Just ask your metric stairs!

1i-k. Accurate distance measurements are not so important here, but the process of measuring is very important. Be alert for these common problem areas, pointing out errors as you notice them.

a. Begin measurements from zero, not from the edge of the ruler.

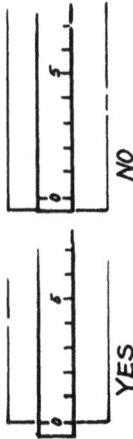

YES   **NO**

b. You *can* start measuring from the high-numbered end of the ruler. But the numbers written on the ruler won't correspond. You must

2. We drew all line segments on Terry's route in whole centiLengths. So you can measure distances equally well with a sweetruler or with a line of sugar cubes. It's easiest to measure each short line segment, then add them all together.

Evaluation

Q: a. Is the top of your paper longer than a deka-paper-clip? Explain.

b. Draw a line that is 5 deci-paper-clips long.

A: a. Depending on the size of the paper clips you provide, the paper should measure less than a deka-paper-clip along its top; perhaps 6.5 paper-clips or .65 deka-paper-clips.

b. Students should draw a line equal to half the length of a paper clip.

Materials

☐ A sweetruler measuring tape from activity 2.
☐ A paper clip to hold the sweetruler (optional).
☐ Activity sheet 1.
☐ Sugar cubes (optional).

DECIMAL FLOWCHART

Which way do you move the decimal point when you divide by 100? How many places do you move it? If your class asks questions like these, they have not mastered the decimal system.

Here is an exercise to provide just the experience they need. Display the empty flowchart (below and to the left) for all to see. You might draw it on your blackboard, project it onto an overhead, or run off individual copies.

Pick a number, any number, to enter in the top input circle. We'll enter the number 1 because it provides lots of practice using zero place holders. Now follow it through all the operations to the bottom output circle. (See below and to the right.)

If you perform each operation correctly you'll end up with the same number you started with. So the flowchart is self-checking. Ask volunteers to enter other inputs.

Do your students need more practice? Have them draw up their own flow charts with their own sequence of operational commands. Then they can trade and try them out on each other.

NAME:

CLASS:

Metric Measuring ( )₃

(TO) practice measuring distance with a ruler. To express each measure in units with two different metric prefixes.

ONE DIMENSION

1 Use your sweetruler to measure each distance.

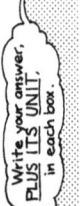

Write your answer, PLUS ITS UNIT, in each box.

Answer in CENTILENGTHS (cL)

a.	3 cL
b.	5 cL
c.	2 cL
d.	1.5 cL
e.	6 cL
f.	4.5 cL
g.	5.2 cL
h.	4.7 cL

Answer in MILLILENGTHS (mL)

30 mL
50 mL
20 mL
15 mL
60 mL
45 mL
52 mL
47 mL

Answer in DECILENGTHS (dL)

varied answers

Answer in LENGTHS (L)

varied answers

i.	The length of your room?
j.	The length of your desk?
k.	The height of the doorway?

2 This map shows the path Terry takes each day to school. Let the side of a sugar cube (1 centiLength) stand for a hectoLength . . . .

SCALE: 1 centiLength (sugar) equals 1 hectoLength (on map)

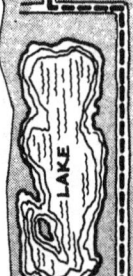

SCHOOL   LAKE   HOME

Answer in HECTOLENGTHS (hL)

a.	11 hL
b.	3 hL
c.	9 hL

a.	How far must Terry walk to school?
b.	How long is the lake?
c.	How far would a crow fly from school to Terry's house?

Answer in KILOLENGTHS (kL)

1.1 kL
.3 kL
.9 kL

Flowchart with Input

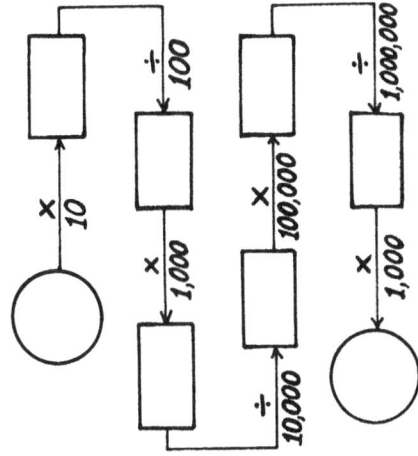

A flowchart:
- 1 → ×10 → 10 → ÷100
- 100 → ×1,000 → .1 → ÷10,000...

```
1  ×10   → 10  ÷100
           .1
100 ×1,000  .01  ÷10,000
            100,000
   ×100,000
1  ×1,000  .001  ÷1,000,000
           1,000,000
```

Empty Flowchart

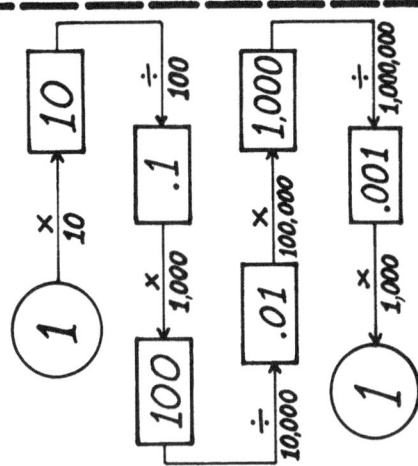

```
○  ×10        □   ÷100
   ×1,000         ÷10,000
□              □
   ×100,000
○  ×1,000     □   ÷1,000,000
```

To verify that each calculated area is correct, cover as much area with whole squares as you can, but leave the fractional spaces open. Then draw lines to show how the fractional squares fit in.

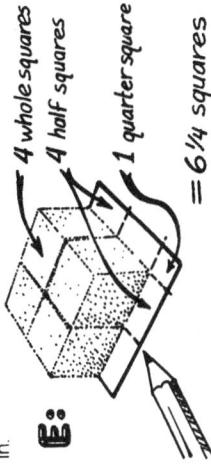

↱ 4 whole squares
↱ 4 half squares
↱ 1 quarter square

= 6¼ squares

Evaluation

Q: Use a paper clip to measure the area of this box in square paper clips.

(Draw a box that measures 2 clips by 3 clips—larger than this scale drawing. Don't draw the paper clips.)

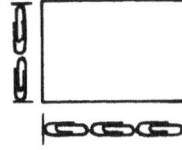

A: A = (2 p.c.) (3 p.c.) = 6 sq p.c.

Materials

☐ A sweetruler from activity 2.
☐ Hand calculator (optional).
☐ Sugar cubes.

TOPS peoplets pop up all over this worksheet reminding students to **write units** with each answer. This is especially important in lessons where you need to distinguish between two or more dimensions, in this case 1-dimensional distance and 2-dimensional area.

When you see numbers written without units refer to them as so many "apples" or "oranges" or something ridiculous. A number by itself is unspecified. It can refer to anything. So call it anything.

Forgetting to write units is a careless habit that students easily fall into. You'll have to work hard to break it. One strategy is to make a firm rule: in this worksheet and all others, a unitless measurement is no measurement at all. Simply refuse to accept unfinished worksheets until they are properly completed—numbers *and* units.

Because the numbers in this table are arranged in columns, some may substitute ditto marks for units written above. This is not good practice. You can easily write ditto marks without thinking—and thus without learning. Making an extra effort to write out the units (they're only abbreviations) is good medicine worth taking.

The areas of boxes A, B, and C are easy to both calculate and verify. You need only multiply whole numbers together. All the squares fit nicely inside with no overlapping.

Not so for boxes D and E. These have half-centiLength dimensions. If multiplying fractions is beyond the ability of your students, you can always supply calculators.

(TO) develop a concrete understanding of area. To learn how to calculate area and express it as squared measure.

NAME:

CLASS:

SQUARED DIMENSIONS

Metric Measuring( )₄

Don't forget the UNITS!

1. Measure both sides of this box, then multiply to find the area.

2. Measure this box and all the others in the same way.

Write measurements AND THEIR UNITS in the table below.

B

E

D

A

C

Double check your math answers — cover each area with sugar lumps and count the squares!

Did you write all the UNITS?

To find AREA just multiply:

(length) × (width) = Area

cL × cL = sq. cL

WRITE THE UNITS

box	length	×	width	=	Area by multiplying	Area by covering with sugar lumps
				=	by multiplying	by covering with sugar lumps
A	5 cL	×	4 cL	=	20 sq cL	20 sq cL
B	6 cL		3 cL		18 sq cL	18 sq cL
C	4 cL		2 cL		8 sq cL	8 sq cL
D	3 cL		2.5 cL		7.5 sq cL	7.5 sq cL
E	2.5 cL		2.5 cL		6.25 sq cL	6.25 sq cL

TOPS LEARNING SYSTEMS

Teaching Notes 5

In the last worksheet we made a big deal about writing units with each measurement. All of our admonitions apply here as well, and in worksheets to come. Unitless measurements don't mean a thing. Challenge any student who disagrees to give you 5. (No, it's not a handshake unless you say it is. And then you've specified the unit!)

Extension

Estimate how many sugar cubes fill your classroom. Is it more than a million? More than a billion? Show all your math.

Evaluation

Q: A box measures 3 paper clips by 4 paper clips by 5 paper clips. Find its volume.

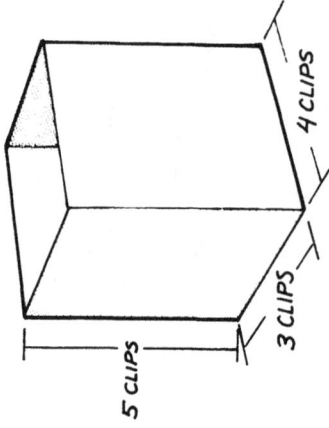

5 CLIPS
3 CLIPS
4 CLIPS

A: V = (3 p.c.) (4 p.c.) (5 p.c.) = 60 cu p.c.

Extend these boxes into 3-dimensional volumes by stacking sugar cubes to the indicated heights. The first 3 volumes in the top row contain only whole cubes, with no parts overlapping outside the lines. So each volume is the simple product of 3 whole numbers. You can easily verify this by counting all the cubes in each stack.

It may be necessary to demonstrate how to find the product of 3 numbers. First you multiply 2 dimensions together. Then multiply your resulting product by the last dimension. When confronted with a row of 3 numbers, some students may confuse operations, taking the sum instead of the product.

The boxes that are stacked 4 high and 5 high are more challenging because they have half-centiLength dimensions. Again, if your class can't multiply fractions together, consider using hand calculators.

To verify that each calculated volume is correct, stack the cubes that fit completely inside the rectangles, but leave the fractional spaces open. Then draw lines to show how the fractional cubes complete the cover. When counting the cubes you'll have to include both real ones and imaginary ones.

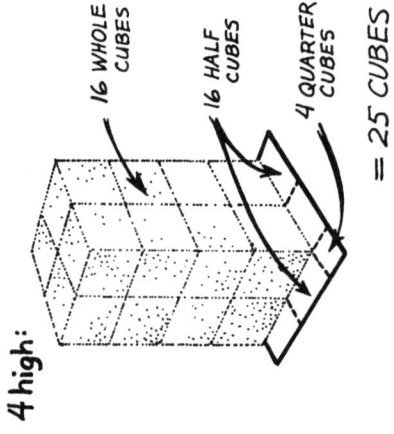

16 WHOLE CUBES
16 HALF CUBES
4 QUARTER CUBES

4 high:

= 25 CUBES

The last small space is filled by just 1 sugar cube. Multiplying together its three unit dimensions yields its unit volume—1 cubic centiLength. To find how many sugar cubes will fill up any space, your classroom for instance, simply find the volume in cubic centiLengths.

Materials

☐ A sweetruler.
☐ Hand calculator (optional).
☐ Sugar cubes.

(TO) develop a concrete understanding of volume. To learn how to calculate volume and express it as cubed measure.

NAME: _____ CLASS: _____

Metric Measuring( )5

CUBED DIMENSIONS

3 High

4 High

1 High

5 High

1. Imagine that this box is covered with sugar cubes stacked **1 High.**

Measure all 3 sides and find the volume.

2. This box is stacked **2 High,** and the others 3 high, 4 high 5 high and 1 high.

Find each measure and complete the table.

Double check your answers — build each volume and count the cubes.

Write measurements in the table below...

Multiply to get AREA, then multiply again to get VOLUME:

(Area) x (height) = Volume

cL x cL x cL = cu. cL

WRITE IN THE UNITS

box	length	X	width	X	height	=	Volume (by multiplying)	(by counting cubes)
1	5 cL	X	4 cL	X	1 cL	=	20 cu cL	20 cu cL
2	4 cL		3 cL		2 cL	=	24 cu cL	24 cu cL
3	3 cL		3 cL		3 cL	=	27 cu cL	27 cu cL
4	2.5 cL		2.5 cL		4 cL	=	25 cu cL	25 cu cL
5	5 cL		.5 cL		5 cL	=	12.5 cucL	12.5 cucL
1	1 cL		1 cL		1 cL	=	1 cu cL	1 cu cL

TOPS LEARNING SYSTEMS

(TO) measure a box in one, two and three dimensions. To estimate quantity by measuring.

Metric Measuring ( )₆

LENGTH...AREA...VOLUME

1 Fold two 3 x 5 index cards exactly in half the short way.

2 Tape the cards together to form a box.

3 Measure this box to the nearest centiLength with sugar cubes. Be sure to *include the correct unit* with each answer.

a. height =	5 cL	
b. length =	4 cL	
c. width =	4 cL	
d. area of side =	20 sq cL	
e. volume =	80 cu cL	
f. area of top =	16 sq cL	
g. length of all 12 edges =	52 cL	
h. surface area = (all 4 sides)	80 sq cL	

4 How many sugar squares fit on the surface of your desk? Show your math.

(Measure the length and width of your particular desk, then multiply to find the number of sq cLs, or sugar squares.)

5 How many sugar cubes could a grocery bag hold? Show your math.

Here are the results for one particular grocery bag:

height = 27 cL
length = 19 cL
width = 11 cL

V = (27cL)(19cL)(11cL)
V = 5643 cu cL or sugar cubes

TOPS LEARNING SYSTEMS

Sugar cubes and 3 X 5 index cards are a perfect match. Fold two cards into a box and you'll find that an even number of cubes fit snugly inside. We use this box to summarize and reinforce concepts of length, area and volume presented in activities 3, 4 and 5. All measurements come in easy-to-compute whole centiLengths.

A great way to introduce this lesson is to physically act out each of the three dimensions. You go through the motions and have your class identify each dimension in unison reply.

Once your students are familiar with each visualization, reverse the procedure. You call out length, area or volume and allow your class to respond kinesthetically. The mind absorbs more rapidly and more completely when you involve the whole body in the learning process.

You call out	*Your class acts out* . . . .
length	length
area	area
volume	volume
side times side	area
area times side	volume
side times side times side	volume
centiLengths	length
square centiLengths	area
cubic centiLengths	volume
1-dimension	length
2-dimensions	area
3-dimensions	volume

This activity is sometimes more therapeutic than a mid-morning break or recess!

4-5. Beyond 10 centiLengths, the sweetrulers are inaccurate. So it's reasonable to round off to the nearest whole centiLength. This will simplify area and volume calculations as well.

It is interesting to compare class answers. Differences in length are small compared to area variation. Differences in volume are the most extreme. This illustrates how little experimental errors quickly multiply into big ones.

Evaluation

Q: This box is marked off in units of measure called "tinys." Find each measure:

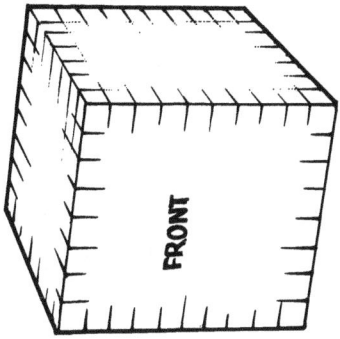

FRONT

height of box = _____
area of front = _____
volume of box = _____

A: height = 10 tinys
area = 80 square tinys
volume = 560 cubic tinys

Materials

☐ 3 X 5 index cards.
☐ Cellophane tape.
☐ Sugar cubes.
☐ A sweetruler.
☐ A large grocery bag.
☐ Hand calculator (optional).

Teaching Notes 6

no name tag. (Don't let your students see you remove it.) Center the balance so everyone can observe that it balances level. Then lift the straw beam above the clothespin, say a few magic words and put it down again. Because you are moving the beam, no one will notice that you twisted it 180° in your fingers so it now rests upside-down. Now no one can balance it again unless you break the magic spell!

14. By pulling the string back up and taping it a second time, you insure that it won't pull loose. It stays firmly attached to the cup even when weighing heavy objects.

16. Each cup has a slightly different mass. Here you associate a particular cup with a particular side.

Worksheet Answers

17. A sugar cube is heavier.

20. A penny weighs between 5 and 7 paper clips. Pre-1982 pennies weigh about 1 paper clip more than post-1982 pennies. See activity 14, step 6.

Evaluation

Q: How do you center a balance?

A: First empty both weighing cups. Then move the tape rider left or right on the beam to a position where it will balance perfectly level.

Materials

☐ Supplementary worksheet. This activity is 2 pages long. Don't forget to hand out the second page.
☐ String. Kite string has about the right thickness.
☐ Scissors.
☐ Plastic soda straws. See teaching notes 18.
☐ Paper clips of uniform size, about this large.

Banish all odd-sized paper clips from your classroom for the duration of this module.
☐ A straight pin. These come in a variety of lengths. Buy extra longs, or the longest available.
☐ A wooden spring-action clothespin.
☐ Cellophane tape.
☐ A standard soda pop can with pull tab attached. Or substitute a soda bottle. See lesson 8, note 3, paragraph 4.
☐ Small styrofoam cups—about 6 ounces. Avoid using large or jumbo sizes. In dry climates a static charge may buildup on styrofoam, causing the weighing cups to cling to the table surface. If this is a problem, substitute paper cups.
☐ A sugar cube.
☐ A penny.

2. Be careful to tie the knot so only a small amount of string hangs outside the loop. If you leave an excess, the loop will not be long enough to poke out both ends of the straw in step 4.

3. To find the center of the straw, lay it on the guide located to the right of the worksheet. Move it back or forth until each end is *equidistant* from the same letter. Then, *without moving the straw*, poke the pin on center. Don't push it clear through yet. A light poke is all that's necessary.

Accurate placement of this pin-poke is extremely important. If both balance arms are not the same length, the shorter arm will always weigh too light and the longer arm too heavy. See teaching notes 11 and 12, A QUESTION OF BIAS.

We provide a teacher's check so you can *personally* verify that each student has poked the straw on center. If you find that it is even slightly off, repoke a more accurate hole. Circle it in pencil to use as the true center for the pin pivot in step 6.

You can eliminate all problems (and part of the challenge) by prepoking each straw yourself. Be careful. Straws from the same box may not have the same length.

5. Because it has been cut open along its entire length, this second straw narrows down to slide easily inside the first. Notice that the string gets sandwiched *between* the inner and outer straws so that it remains locked in a fixed position. Students who skip step 4 may incorrectly push the string through the center of *both* straws, where it can easily shift about.

8. The beam is most stable when you rest the pin on the solid wood (not on the tape). It may be necessary to scrape off the last traces of tape with the edge of your scissors.

9. Tell your students that the beam will not likely balance level until step 15. Otherwise they will worry that it's not working properly.

This activity continues on a supplementary worksheet. Please refer to it when interpreting the rest of these teaching notes.

10. 15. Putting your name at the top of the beam not only identifies the owner, but also establishes an up-down position for the beam. This is important. In step 15, if you turn the beam over so the string loops emerge from the top of the straw instead of the bottom, it simply won't balance. That's because the beam's center of gravity is raised higher than the pivot. Students who complain their balances are acting weird have likely turned them upside-down.

To emphasize this point in a dramatic way, try doing this magic act. Start with a balance that has

---

NAME: _____

CLASS: _____

Metric Measuring ( )7

BUILD A STRAW BALANCE

(TO) construct an equal-arm, soda-straw balance to use in all weighing activities in this module.

1 Double some string to make a long U-shape.

Cut its length equal to 1 straw plus 1 paper clip.

2 Tie the string in a loop. Keep the knot at one end.

KEEP THESE ENDS SHORT.

3 Lay your straw on the one shown at right so each end is the exact same distance from the same letter . . .

. . . Lightly poke the exact center and show it to your teacher.

SAME LETTER
MARK CENTER

☐ Teacher Check

4 Push your string loop through your straw so it hangs out both ends.

Keep the knot at one end.

KEEP THE KNOT HANGING OUT.

5 Cut open a new straw along its full length . . . .

. . . Then slide it inside your first so the string is pressed *between* both straws.

6 Push a pin *straight* through the middle of the straw where you poked it before. Show it to your teacher.

HOLD ONTO THE STRING—DON'T LET IT SLIP INSIDE.

PUSH STRAIGHT THROUGH — NOT CROOKED

☐ Teacher Check

7 Fold tape over the ends of a clothespin. Each piece should stick out past the end about as wide as a paper clip.

PINCH TAPE FLAT

8 Cut out a narrow strip from the center of the tape.

LOOKS LIKE EARS!

CUT TO THE WOOD

9 Clamp the clothespin to the pull-tab on a pop can.

REST PIN IN SLOTS

BE SURE THE LOOPS HANG OUT THE BOTTOM... ...TURN THE STRAW OVER, IF THEY DON'T.

I H G F E D C B A   A B C D E F G H I

TOPS LEARNING SYSTEMS

Activity Sheet

(TO) invent a unit of measure that corresponds to grams in the metric system.
To learn to use the soda-straw balance.

NAME: _____ CLASS: _____ Metric Measuring ( )8

SWEET MASSES

I hereby name thee ONE MASS!

1 Let's say a sugar cube weighs one "MASS".

2 If a sugar cube weighs 1 Mass, how many cubes weigh a kiloMass? Complete the table.

PREFIX:	MEANING:	UNIT:	HOW MANY SUGAR CUBES:
kilo	thousand	kiloMass	1000 sugar cubes
hecto	hundred	hectoMass	100 sugar cubes
deca	ten	decaMass	10 sugar cubes
	—	Mass	1 sugar cube
deci	tenth	deciMass	.1 sugar cubes
centi	hundredth	centiMass	.01 sugar cubes
milli	thousandth	milliMass	.001 sugar cubes

3 Answer each question using your balance. Remember to ...

... keep your name tag UP and the strings DOWN...

... center your empty balance with the tape tab each time before you weigh.

a. How many paper clips weigh 1 Mass?

b. How many pennies weigh 1 dekaMass?

c. How many rice grains balance a paper clip?

= one Mass

= one deka-Mass

= one

Your answers may vary from these.

7½ paper clips   *14 pennies*   *31 rice grains*

d. You know how how many rice grains balance a paper clip and how many paper clips weigh 1 Mass. Calculate how many rice grains weigh 1 Mass.

e. Is a rice grain larger or smaller than a milliMass? Explain how you know.

1 Mass = 233 rice grains = 1000 milliMass

Since it takes many more milliMasses to make a Mass than rice grains, a rice grain must be LARGER.

$$\frac{31 \text{ rice grains}}{\text{paper clips}} \times \frac{7.5 \text{ paper clips}}{\text{Mass}} = 233 \text{ rice grains / Mass}$$

TOPS LEARNING SYSTEMS

Teaching Notes

openings with the right diameter, so a clothes-pin will fit snugly inside. Others may have a mouth that is too large. You can make it narrower by sticking tape around the inside of the rim.

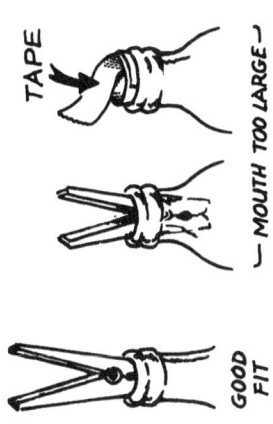

TAPE — GOOD FIT — MOUTH TOO LARGE

Language is not always accurate in a technical sense. We say the sun "sets" even though we know it doesn't. All over the world people "weigh" things in "grams," even though grams are units of mass, not weight.

We don't think your students will be mislead by this kind of talk. When they are intellectually ready, they will have no trouble learning that the earth spins on its axis, and that weight is the product of mass times acceleration. So in this module, we have decided not to stress the distinction between mass and weight. To write clear, easy-to-understand directions, we will use mass and weight interchangeably, as in normal English usage.

As long as we confine our measurements to planet earth at elevations reasonably near sea level, this is O.K. The relationship between weight and mass only changes as you change the acceleration (gravitational pull) acting on a body.

Consider a candy bar. On earth we say it "weighs" 100 grams and get away with it. But in free space—on an orbiting space station for example—this 100 gram mass weighs nothing at all. Weight, a force, changes with the acceleration of earth's gravity according to Newton's famous formulation, $F=ma$. Fortunately for space travelers, the candy bar's mass (all those peanuts, the nougat and milk chocolate) remains constant.

3e. Those who are unable to find the number of milliMasses in a Mass should consult their metric stairs in activity 1. Begin with 1 Mass, then step down 3 steps (move the decimal right 3 places) to 1000 milliMasses.

You can also arrive at this result algebraically. A milliMass equals .001 Mass. So you need to collect 1000 of these tiny weights to balance 1 Mass:

1 milliMass = .001 Mass
1000 (1 milliMass) = 1000 (.001 Mass)
1000 milliMass = 1 Mass

1. By defining 1 sugar cube to weigh "1 Mass", we continue to build a sweet measure system identical in structure to the metric system. This 1-sugar-cube "Mass" parallels a metric gram just as a 100-sugar-cube "Length" corresponds to a meter.

3. Any balance can easily drift off-center with routine handling. Bits of dirt, sugar or other debris can also make the beam tilt this way or that. So you should check for centeredness before each new weighing. If the beam is out of balance, it only takes a moment to move the tape tab a little to the right or left.

If the balance won't center at all, the beam is likely turned upside-down. Check to see that the name tag points up and that both string loops point down. To balance, the weighing cups must pull on the beam at points that are lower than the pivot pin.

A good way to increase the overall stability of the balance base is to fill the can 1/3 full with gravel, sand or even dirt. You can also hold the clothespin more tightly to the tab by winding it with tape.

You can substitute soda bottles for pop cans to serve as balance bases. Some bottles may have

Designate an out-of-the-way place where your students can set their balances aside until activity 13. They may be reluctant to give them up so soon. So be upbeat. When they do use their balances again, they'll develop a complete set of gram weights, then do a whole series of creative mass experiments. It's something to look forward to.

Evaluation

Q: A nickel weighs about 10 paper clips.
a. How many paper clips does a kilonickel weigh?
b. How many paper clips does a decinickel weigh?

A: a. About 10,000 paper clips.
b. About 1 paper clip.

Materials

☐ Activity sheet 1. Students may need to use the metric stairs as a reference.
☐ A soda straw balance from activity 7.
☐ Paper clips.
☐ Sugar cubes.
☐ Pennies.
☐ Uncooked rice grains. We used white long-grained rice.

Teaching Notes 8

Teaching Notes 9

There are 2 ways to find the volume of a container. You can measure the various dimensions with a ruler, then mulitply. (This is called dry measure.) Or you can fill up the container with measured amounts of water. (This is called liquid measure.)

In this activity we define our unit of liquid measure, "1 Volume", to be the space occupied by 1000 sugar cubes. This corresponds to metric liquid measure based on 1 liter.

With these three units of sweetmeasure now in place—Lengths, Masses and Volumes—get set for the great metric conversion. Its coming up in the next activity!

1. It's easy to cut well-formed squares, 10 centiLengths to a side, when you use notebook paper. If you let one corner of your square serve as one corner of your square, as illustrated, you only need to make two cuts—one parallel with the lines in the paper, and the other perpendicular to these lines.

4. The sweetbox measures 10 cL on each side. This multiplies out to a dry measure volume of 1000 cubic centiLengths. Recall that 1 sugar cube equals 1 cu cL (see activity 5). So exactly 1000 sugar cubes fit inside the box.

Another way to understand this is to stack 10 sugar cubes along all three dimensions. Imagine 10 rows of 10 cubes in a layer (that makes 100) stacked 10 layers high (that makes 1000).

6. This cube pattern folds to a measuring cup that will actually hold water. By covering it with tape, you make it waterproof, at least long enough to complete the experiment.

7. The box will be better formed if you first fold each flap over to form a sharp crease along the bottom edge. This places the flaps in a near-upright position where they can be taped at vertical right angles.

If some students have trouble keeping the flaps in position, suggest they place a sugar cube in the middle. This keeps the sides from being pushed too far inward.

others, the experimental process is more important than any particular answer. Do your students understand how water is used to measure volume? This is most important.

9. The sweetbox, of course, won't hold water. To arrive at an answer, you'll have to first find how many milliVolumes fill a styrofoam cup, then divide your answer into 1000 milliVolumes—the capacity of the sweetbox.

To find the volume of a styrofoam cup, it's easiest to fill a second cup 10 mV at a time with the one just calibrated in step 8. Some students may fill it instead with their small cubes, one mV at a time.

Either way, after they determine the volume of their cup, they still have to use division to find the number of cups in their 1000 mV sweetbox. Some may entirely forget to do this second part of step 9.

10. The small 1 mV cube measures 1 cL by 1 cL by 1 cL. This makes 1 cubic centiLength. That's an important result. Liquid milliVolumes equal dry cubic centiLengths, unit for unit.

That's how it is in the metric system too. Liquid milliliters equal dry cubic centimeters. But let's not get ahead of ourselves. These are concepts for lesson 16.

Evaluation

A: a. There are 1000 millispoons in 1 level teaspoon. Five level teaspoons fill a pill box. Six pill boxes fill a styrofoam cup. How many millispoons fill a styrofoam cup?

1000 millispoons

b. Which has greater volume, a styrofoam cup or a hectospoon?

A: a. 30,000 millispoons.
b. A hectospoon. It has a capacity of 100 teaspoons. A styrofoam cup holds only 30 teaspoons.

Materials

☐ Lined notebook paper. Sheets that have square corners are better than paper with rounded corners; either kind will do.
☐ A sweetruler.
☐ Scissors.
☐ Cellophane tape.
☐ Styrofoam cups.
☐ A sink with running water, or a tub partly filled with water.

NAME:

CLASS:

Metric Measuring ( )₉

SWEET VOLUMES

(TO) invent a unit of measure that corresponds to liters in the metric system. To distinguish liquid measure from dry measure.

1 Cut 4 squares from lined notebook paper.

Make each square 10 centiLengths on each side . . .

2 Tape the 4 squares together to form a "sweetbox".

3 Let's call the space occupied by your sweetbox 1 VOLUME.

Label your box this way.

4 How many sugar cubes fill your sweetbox? **1000 sugar cubes**

5 Complete this table.

1 sweetbox	1 Volume
.1 sweetbox	1 deciVolume
.01 sweetbox	*1 centiVolume*
.001 sweetbox	*1 milliVolume*

1 Volume	1000
	100
	10
	1

1 Volume =	1000 milliVolumes

sugar cubes
sugar cubes
sugar cubes
sugar cubes

6 Cover this milliVolume box pattern completely with tape to make it waterproof. Then cut it out.

OVERLAP THE TAPE STRIPS.

1 mV

7 Fold the flaps so the tape is on the inside of the box.

8 Pour 10 cubes of water into a styrofoam cup. Mark the 10 mV level on the inside.

10 mV

Seal all 4 edges with tape.

9 How many *cups* fill your sweetbox? Show your work.

FiRST find how many mV's are in one FULL cup!

1 full cup = 44 mV

$$\frac{1\,cup}{44\,mV} \times \frac{1000\,mV}{sweetbox}$$

= 23 cups / sweetbox.

(This answer is based on 6 1/3 oz. strofoam cups.)

STYROFOAM CUP

10 You can measure volume in 2 ways:

1 USE A RULER . . . . . . this is called DRY MEASURE

How many *cubic centiLengths* in . . .
a sugar cube? **1 cu cL**
a sweetbox? **1000 cu cL**

2 USE A MEASURING CUP or CUBE . . . this is called LIQUID MEASURE

How many *milliVolumes* in . . .
a sugar cube? **1 mV**
a sweetbox? **1000 mV**

TOPS LEARNING SYSTEMS

(TO) construct paper models of a gram, a meter and a liter based on one cubic centimeter of water.

Discussion

Compare each meter tape to a commercial meter stick. Find out who made theirs the most accurate. Discuss sources of error.

Extension

In 1983 a world conference on weights and measures redefined the meter in terms of the speed of light, a physical constant known with great accuracy. One meter is now internationally accepted as the distance light travels in 1/299,792,458 of a second. (That's precisely 100 metric water cubes placed end to end, of course!)

How did scientists decide the length of a meter before they measured the speed of light? Send selected students to the library to research the meter's history.

Evaluation

Q: a. How can you distinguish a metric water cube from a non-metric one? Give a complete description.

b. How many _____ metric cubes end to end reach 1 meter? _____ metric cubes fill a liter? _____ metric cubes weigh a kilogram?

A: a. A metric water cube measures 1 centimeter on each side. It has a volume of 1 milliliter (or 1 cubic centimeter) and a mass of 1 gram.

b. 100 cubes reach 1 meter.
1000 cubes fill a liter.
1000 cubes weigh a kilogram.

Materials

☐ Sugar cubes (optional).
☐ Scissors.
☐ Cellophane tape.
☐ Adding machine tape or strips of paper. Precut to roughly 1.5 meters or 5 feet long. You can use your own height or arm span to make a rough estimate. Your students will trim off the excess. As before, if adding machine tape is not available, try substituting a roll of continuous paper towel that you can write on. You or your students will have to cut off long narrow strips from this roll, about 6 metric water cubes wide.
☐ Lined notebook paper.
☐ An index card (optional). This can be used as a straight edge to draw lines for the liter box in step 3.
☐ A meter stick (optional). You may want to compare student-made meter tapes to this standard. See the discussion above.

Teaching Notes 10

Grams, meters and liters use water cubes, not sugar cubes. Because a metric water cube is somewhat smaller than a sugar cube, the gram is lighter than a Mass; the meter is shorter than a Length; the liter is smaller than a Volume. But not by much. The units in both systems have the same order of magnitude. And they follow the same systematic logic. So put away those sugar cubes. Build a metric water cube instead. Converting to metrics is that easy!

Do you want to generate lots of metric enthusiasm? Try a few rounds of goodby-hello.

You say	_Your class responds . . . ._
Goodby Mass.	Hello gram.
Goodby Length.	Hello meter.
Goodby Volume.	Hello liter.
Goodby centiLength.	Hello centimeter.
Goodby kiloMass.	Hello kilogram.
Goodby milliVolume.	Hello milliliter.
Goodby cubic centiLength.	Hello cubic centimeter.

Keep it up as long as the enthusiasm lasts.

1. This little metric water cube, modest though it may be, is the logical spring from which the entire metric system flows. Your students will refer to it often throughout the remainder of this module.

Because it is so small yet so important, encourage your students to cut, fold and tape the cube together using extra care and patience. They will do their best job if they fold the edges _first_, then stick it together with _small_ slices of cellophane tape.

2. Every time you mark the width of the water cube, then move it up to an adjacent position, you accumulate measuring error. So don't expect to produce a standard meter. Our purpose here is not to make an accurate measuring tape; rather to reinforce the concept that 1 meter measures 100 metric water cubes (centimeters) placed end to end.

3. Make this liter box as you made the sweetbox in Activity 9. If you measure accurately, you can cut all 4 sides of the box from a single sheet of notebook paper.

In practice, most liter boxes turn out too large to do this, for the same reason that the meter tapes turn out too long.

4. Ask your students to throw this metric cube as often as necessary until they have thoroughly memorized all its dimensions. Reinforce this learning with a class review. You call out one side of the cube and have your students chorus back the opposite side. Don't forget to ask them its volume as well—1 cu cm.

NAME: _____   CLASS: _____

Metric Measuring (   )₁₀

SUGAR CUBES/WATER CUBES

SWEET measure works just like METRIC measure.

... METRIC measure is based on a WATER cube!

Sweet measure is based on a SUGAR cube ...

SUGAR CUBE: WATER CUBE:

— Actual Size —

To measure in METRICS we must replace sugar cubes with water cubes, and learn just

3 NEW WORDS

1 Mass is in GRAMS.

a. A Mass equals 1 sugar cube. So how heavy is a gram?

1 water cube

b. Cut out this metric water cube pattern. Fold and tape it together using _tiny_ pieces of tape.

This cube filled with WATER weighs **1 GRAM**.

2 Length is in METERS.

a. A Length (sweetruler) equals 100 sugar cubes. So how long is a meter?

100 water cubes

b. Use your metric water cube to make a strip of paper 1 meter long. Number each 10 spaces and label it "1 Meter".

Each unit is one metric water cube 1 metric cm long!

LABEL   1 METER

METRIC WATER CUBE

3 Volume is in LITERS.

a. A Volume (sweetbox) equals 1000 sugar cubes stacked together. So how big is a liter?

1000 water cubes

b. Use your metric water cube and lined notebook paper to make a liter cube. Label it "1 Liter".

1000 WATER CUBES

A cube 10 times longer is 1000 times bigger!

10 WATER CUBES

4 Memorize the length, mass and volume units of your metric water cube ...

OPPOSITE SIDES : GO TOGETHER

... Then toss the cube. Read the top to predict the bottom.

Mark ☑ if you got it right; ☒ if you got it wrong ....

20 TOSSES:

SAVE YOUR 3 METRIC MODELS

DPS LEARNING SYSTEMS

Class Discussion

Almost everything you need to know about the metric system is summarized on this single worksheet! Prefixes, decimal conversion factors, common metric equivalents—it's all here. And the metric models from activity 10 tie it all together.

Remember the cooperative game of metric dice we first introduced back in activity 1, step 7? A fun way to ensure a positive learning experience in this worksheet is to have everyone play a game or two before proceeding.

Does your class need extra help on this worksheet? Try this demonstration. First present a meter tape that someone made in activity 10. Start at the top of the metric stairs and ask, "How many of these meter tapes make a kilometer?" Work on down the stairs through hectometers, dekameters... millimeters. (You'll need to divide one of the centimeter divisions into 10 parts to demonstrate a millimeter.)

Next, walk the liter stairs. Ask how many milliliters make 1 liter. (Start with 1 liter, then multiply down 3 places to 1000 ml.) Confirm this relationship with your metric water cube. Imagine 10 rows of 10 cubes in a layer, stacked 10 layers high. That makes 1000 water cubes, or 1000 ml.

Evaluation

Q: Fill in each box with equal measure.

1 meter = ▭ cm	1 gram = ▭ kg
1 meter = ▭ mm	1 gram = ▭ mg
1 meter = ▭ km	1 liter = ▭ ml

A:

1 meter = 100 cm	1 gram = .001 kg
1 meter = 1000 mm	1 gram = 1000 mg
1 meter = .001 km	1 liter = 1000 ml

Materials

☐ All metric models from activity 10—the metric water cube, the meter tape and the liter box.

Worksheet Answers

3. These metric units are not drawn to scale.

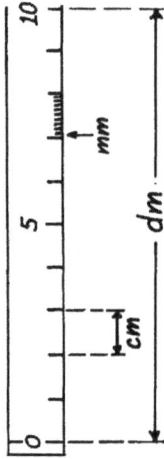

4. 1 liter = 1000 ml.
   1 liter water = 1000 g or 1 kg.

A QUESTION OF BIAS

Here is a demonstration that explains why unequal-arm balances don't weigh true. Use it any time you wish. We think it fits in best just before activity 13.

Prepare an unequal-arm balance in advance. Construct it in the usual manner, per instructions from activity 7. Instead of putting the pin on center, move it ½ centimeter (⅜ inch) to the left. Compensate for the extra weight on the longer right arm by taping 2 or 3 pins, out of sight, to the bottom of the left cup. Center the unequal-arm balance with a tape rider as usual. It should look something like this:

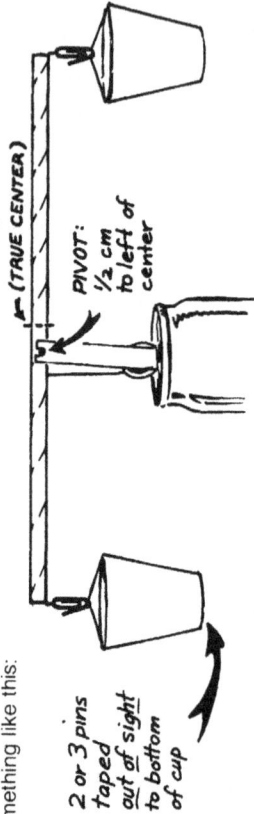

Next show the balance to your class. Adjust the tape rider so it balances perfectly level for all to see. Give no hint that anything is wrong with it. Because the pivot is only slightly off-center, it should appear normal to the casual observer.

Challenge your class to give you two items that have equal mass. As long as you test what they give you on this "special" balance, all equal masses will appear to weigh differently.

(TO) list the metric units in order. To express the most frequently used measures as decimal equivalents.

Metric Measuring ( )11

METRIC STAIRS (2)

1 Write the correct metric unit on each stairstep. Don't abbreviate.

Remember: kilo-, hecto-, deka-, deci-, centi-, milli-

Units with stars are used often. Learn them!

- 1000 meters — **kilometer** ★
- 100 meters — **hectometer**
- 10 meters — **deckameter**
- 1 meter — **meter** ★
- 0.1 meter — **decimeter**
- 0.01 meter — **centimeter** ★
- 0.001 meter — **millimeter** ★

- 1000 grams — **kilogram** ★
- 100 grams — **hectogram**
- 10 grams — **deckagram**
- 1 gram — **gram** ★
- 0.1 gram — **decigram**
- 0.01 gram — **centigram**
- 0.001 gram — **milligram** ★

- 1000 liters — **kiloliter**
- 100 liters — **hectoliter**
- 10 liters — **deckaliter**
- **liter** ★
- 0.1 liters — **deciliter**
- 0.01 liters — **centiliter**
- 0.001 liters — **milliliter** ★

2 Write equal measures in each box below. Use the decimal stairs to help you.

Stand on the first "step" given in each box
MULTIPLY as you step DOWN (move decimal right)
DIVIDE as you step UP (move decimal left)

1 meter =	1 kilometer =	1 centimeter =
0.001 km	0.00001 km	0.00001 km
100 cm	1,000 m	0.01 m
1,000 mm	100,000 cm	10 mm
	1,000,000 mm	

1 gram =	1 kilogram =	1 milligram =
0.001 kg	0.000001 kg	0.000001 kg
1,000 mg	1,000 g	.001 g
	1,000,000 mg	

1 millimeter =	1 liter =	1 milliliter =
0.000001 km	1,000 ml	0.001 l
0.001 m		
0.1 cm		

3 On your meter tape, draw and label a line equal to ...

... 1 decimeter;
... 1 centimeter;
... 1 millimeter.

4 On your liter model ...

... Write how many milliliters it holds.

... If it were filled with water, write how much it would weigh.

(TO) express length, area and volume in units of measure that all derive from one meter. To relate dry cubic centimeters to liquid milliliters.

NAME:

CLASS:

Metric Measuring( )₁₂

DRY MEASURE

1 Use the metric ruler on the side of this paper to cut a string 1 meter long.

Measure the length of your room in meters.

varied answers

HOW MANY STRING LENGTHS?

2 Cut string to the correct length and tape it to each box.

centimeter:

decimeter:

millimeter:

3 Now cut out your metric ruler and find each measure on this box. Write numbers *and* units.

FRONT SIDE width height length

length = **6 cm**
width = **5 cm**
height = **3 cm**
area of front = **15 sq cm**
area of side = **18 sq cm**
volume = **90 cu cm**

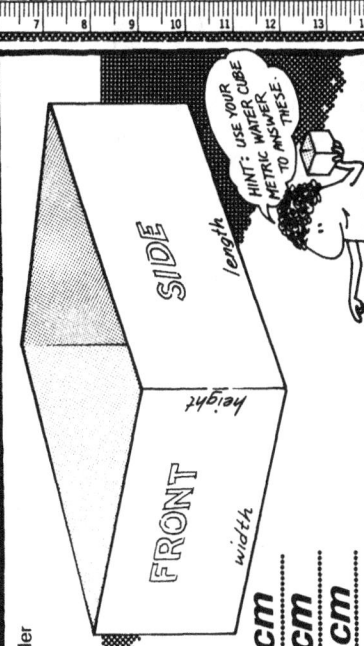

How many milliliters of water fill this box?
How do you know?

**90 ml; because
1 cu cm = 1 ml.**

When filled with water, how much does this box weigh? How do you know?

**90 g; because
1 ml water = 1 g.**

HINT: USE YOUR METRIC WATER CUBE TO ANSWER THESE.

4 Imagine having 11 of these boxes filled evenly to the top with gold dust!
Would you trade them for a liter of gold dust?

(11) (90 ml) = 990 ml

**Yes, trade for 1 liter of gold dust. It contains 1000 ml,
10 ml more than 11 of the above boxes.**

SHOW YOUR MATH!

METRIC RULER 0 cm 1 2 3 4 5 6 7 8 9 10 11 12 13 14 15 16 17 18 19 20

SAVE YOUR RULER

TOPS LEARNING SYSTEMS

Students may bring you 2 coins, 2 sugar cubes, 2 paper clips or 2 identical pieces of notebook paper. When you put any pair in your weighing cups, the longer arm always tilts down.

Some classes may figure out what is wrong with the balance right away. Other classes may require additional prompting. First put identical items, coins for example, in each cup to find out which one is "heavier". Then place each coin in the opposite cup. Now the "heavier" coin is suddenly the "lighter" one. How can this be? Someone will finally notice that the balance always tilts in the same direction, favoring a longer arm.

After the mystery is solved, ask your class why it is appropriate to call your balance "biased". (Because it always favors the same side—the side with the longer arm.)

Your students may want to check their own balances for possible bias. Two identical coins make good test weights. First center the balance. Then place one coin in each cup and observe the tilt (if any). Switch coins to opposite cups and observe the tilt again (if any).

↳ Your balance is biased (off center) if the beam tilts in the ***same*** direction both times. (If this tilt is significant, reposition the pivot pin more on center.)

↳ Your balance is true (has equal arms) if the beam remains level both times.

↳ Your balance is also true if the beam tilts first one way, then the opposite way by the same amount. (The coins don't quite weigh the same.)

The reduced student worksheet to the left has a reduced metric ruler as well. It no longer measures in standard centimeters. But what about the full-sized ruler on the full-sized worksheet? Is this an accurate full scale reproduction of 20 standard centimeters?

Our answer is qualified: yes, as near as we can make it. The ruler is certainly accurate enough to use successfully with this module. But it may not be as accurate as commercially available metric rulers.

We certainly started with a 1-to-1 scale drawing of standard centimeters. But our drawing was then photographed onto a negative, burned into a printing plate, and printed on a press. Then you took the printed worksheet and reproduced copies for your students. As the image was transferred from one medium to another, it may have grown larger or smaller.

Your own photocopier probably introduced the greatest image-size change, perhaps lengthening or shortening your printed original in 1 dimension without a proportional size change in the other dimension. If your copied centimeter ruler doesn't quite measure the copied box in exact whole numbers as we intended, just ask your class to round off to the nearest whole centimeter.

the metric water cube: 1 cu cm (dry measure) equals 1 ml (liquid measure) equals 1 g (for water only). So 90 cu cm = 90 ml = 90 grams.

These equivalents apply specifically to pure water under 1 atmosphere of pressure at its maximum density (about 4°C). But this is being really picky. The equations are true enough for chlorinated tap water at classroom temperatures in mile-high Denver.

4. This problem requires 2-step logic. First find how many milliliters are in 11 boxes. Then compare your answer with 1000 ml, the volume of 1 liter.

Evaluation

Q: This is a metric ruler.

cm 1 2 3 4 5

a. Find the length of a paper clip. . .
   in cm _____
   in mm _____

b. How many of these small rulers. . .
   make a decimeter? _____
   make a meter? _____
   make a kilometer? _____

A: a. A standard-sized paper clip measures about 3.2 cm or 32 mm. The size of your paper clips may be different.

b. A decimeter is 2 rulers.
   A meter is 20 rulers.
   A kilometer is 20,000 rulers.

3. Real rectangular containers have equal opposite sides. But this one isn't real: it's a 3-dimensional projection on flat paper. The length of some lines had to be altered somewhat to make it look like a respectable box. So measure only the dimensions that are labeled.

The last two answer boxes in this step reinforce important metric relationships derived from

Materials

☐ String.
☐ Scissors.

Teaching Notes 12

Dust off your straw balances. They form the centerpiece for all remaining activities in this module. Then turn back to **A QUESTION OF BIAS**, located between Teaching Notes 11 and 12. This demonstration will help your class understand why balance arms must have equal length—as equal as you can possibly make them.

1-4. In these steps you counterbalance 5 folded sheets of paper against a 20 gram mass. Then cut parallel strips off the sheets until they balance level. This yields 5 uniform strips of paper that weigh 4 grams each.

Be sure to evaluate this procedure yourself before your class tries it. If you think your students lack sufficient coordination to do an accurate job, consider this alternative:

You experimentally determine how many centimeters need to be cut off the scratch paper you are using. Standard 8½ X 11, 20-pound-bond—a paper used in many photocopiers—needs slightly more than 3 centimeters shaved off the edge to equal 20 grams. Your particular paper may require a different trim width.

Write your particular result on the blackboard. Have your class trim their sheets of paper to your specifications. (They can use their centimeter rulers from activity 12 for measuring.) Then proceed directly to step 5.

1. Unless the U.S. government at some future date again changes the weight of pennies (see activity 14, step 6), 8 pennies minted in 1983 or later will weigh very close to 20 grams. If U.S. pennies are hard to find in your locality, or if you want to be more organic, try substituting 68 pinto beans or 155 pop corn seeds. Anything at all with a known mass of 20 grams is appropriate.

The most logical way to measure 20 grams is to pour 20 ml of water into your weighing cup. This is the method we recommend if your lab is equipped with graduated cylinders and your students know how to accurately use them.

2. We have previously suggested that you fill the can 1/3 full with sand, gravel or even dirt to increase the stability of the balance base. Because the paper is heavy and awkward, this is especially helpful here. Without this lower center of gravity, the balance easily topples over.

Be sure your class follows all directions in this step very carefully: The sheets must be folded together (like a book) along their length (not width). Slip the *folded* side into the paper clip (so you can trim the paper evenly). Lean the cup out of the way of the paper (don't take it off the beam).

3. If 5 sheets of paper weigh less than 20 grams,

don't proceed any further. Find a heavier grade of paper and start over.

5-7. Each of the numbered quarters weigh about 1 gram. To make a 10 gram weight fold 10 quarters together; to make a 5 gram weight fold 5 quarters together and so on. The 5 and 10 gram weights are bulky, so they need to be held closed with tape. Since tape has weight, use as little as possible.

8. The plastic sandwich bag forms part of the 20 gram mass. So it belongs in the weighing cup along with the salt. If some of the salt misses the sandwich bag and falls directly in the cup, you can add it to the bag later, after you achieve a balance.

9. Don't expect perfection. The beam may tilt a little right or left because of measuring error. It's hard to cut the gram quarters in step 4 perfectly equal.

Evaluation

Q: You have 6 different gram masses: 20g, 10g, 5g, 2g, 2g, and 1g. What combination of these masses would you use to equal . . .

    8g
    17g
    29g
    31g
    36g

A: 8g = 5g + 2g + 1g
    17g = 10g + 5g + 2g
    29g = 20g + 5g + 2g + 2g
    31g = 20g + 10g + 1g
    36g = 20g + 10g + 5g + 1g

Materials

☐ A soda straw balance from activity 7.
☐ New U.S. pennies minted after 1982 or other equivalent weights. See teaching notes 1 above.
☐ Uniform-sized 8½ X 11 sheets of paper of average weight. Medium weight copy paper (20 pound bond) or uniform sheets of scratch paper (one side should be clean) will work fine. To find out if your paper is heavy enough, complete steps 1-3 in the worksheet. Five sheets of your paper must outweigh 8 new pennies (20 grams). Lined notebook paper and very light copy paper (16 pound bond) don't work; they aren't heavy enough.
☐ Scissors.
☐ Cellophane tape.
☐ Plastic sandwich bags.
☐ Table salt.
☐ A spoon (optional). This enables students to more easily control the amount of salt they pour into their weighing cups.

NAME: _____

CLASS: _____

Metric Measuring ( )13

(TO) develop a series of gram paper weights to use with the soda-straw balance.

GRAM QUARTERS

1 Center your balance, then add 20 grams to your left cup.

(20 g = 8 new U.S. pennies)

2 Fold 5 sheets of paper together the long way. Slip the *folded* side under the *right* paper clip so the cup leans out of the way.

FOLD TOGETHER

5 SHEETS

CUP

3 Is this paper heavier than 20 g?

Yes. The beam tilts towards the paper.

4 Cut *parallel* strips off the 5 folded papers until they exactly balance 20 g. Cut only *towards* the fold.

FOLD

CUT JUST A LITTLE AT A TIME . . .

UNTIL IT BALANCES LEVEL . . .

5 Fold each paper into 4 *equal* quarters. Number them from 1 to 20. How much does each numbered quarter weigh?

GATHER 10 QUARTERS

6 Fold 10 g of paper (10 quarters) as small as you can . . .

Tape it together and label it.

10 g

USE VERY LITTLE TAPE

7 Make these other weights with the rest of your paper quarters.

GATHER THESE QUARTERS

5g 2g 2g 1g

FOLD AND LABEL

Don't tape the small ones

5g 2g 2g 1g

1 gram

8 Make a 20 g weight by pouring salt into a sandwich bag.

Twist the plastic closed and tie it in a knot.

FOLD TIGHTLY

10g

20g

Sandwich Bag

Paper Weights

9 Check your weights against each other.

☐ Does 2g balance 2g?

☐ Does 5g balance 2g+2g+1g?

☐ Does 10g balance 5g+2g+2g+1g?

☐ Does 20g balance 10g+5g+2g+2g+1g?

SAVE YOUR WEIGHTS !

TOPS LEARNING SYSTEMS

✓ Add the largest possible mass first, then the next largest and so on. Work down through the lighter masses, in order, until you find the right combination to make the beam balance level. The smaller weights must be reserved to make final small adjustments.

✓ Find the sum of all the weights in the weighing cup. Be sure to keep the decimal points lined up.

6. In 1982 the U.S. government began minting a lighter penny, one that contained much more aluminum and much less copper. During this same year heavier pennies were also produced according to the old formula. As a consequence, 1982 is a year of mixed pennies—both light and heavy. You can avoid all confusion by avoiding 1982 pennies.

You've made a collection of masses suitable for weighing objects from 1 gram to 40 grams. But your balance is sensitive to much smaller masses than these. In this activity, you'll round out your collection by developing a series of smaller masses that weigh only fractions of a gram.

1-2. Instead of counting large gram-sized squares of paper, here you count much smaller graph squares. Even so, the underlying assumption remains unchanged: paper weight is proportional to its area. Thus, if 100 squares weigh 1 gram, then 50 squares weigh .5 gram, 10 squares weigh .1 gram, and so on.

Depending on the kind of paper you use, another number of graph squares (not as tidy as 100) will weigh 1 gram. The answers we suggest in the worksheet are only valid for the photocopy paper *we've* used. If you find that another number of graph squares weigh 1 gram, just divide and multiply that number through the flow chart to find all the fractional gram weights.

Watch out for students who cut out individual squares, then add them to their balance one at a time. This is slow and tedious; it's difficult to count all the squares and you can't use them to make fractional gram weights when they are all cut up into tiny pieces.

3. Remember to label each mass in "grams", not "squares". Never tape these small weights closed. Any additional tape makes them too heavy for accurate use.

4. Because they are cut from uniform graph squares, the equal masses should balance reasonably well. If there is an exaggerated tilt, first check your balance for centeredness. Then recheck your arithmetic and count squares to make sure each weight contains the correct number. If these remedies fail, test your balance for bias. See the last part of A QUESTION OF BIAS between activities 11 and 12.

5. Before your class begins to weigh coins, demonstrate proper weighing procedures. Start with any small object—a clothespin will serve. Find its mass while your whole class looks on. Emphasize these points:

✓ Begin by centering the beam to a level position. Move the tape tab along the straw until it balances.

✓ Always add gram weights to the same cup. Since the right hand is nearest the right cup, right-handers find it easiest to add weights to this cup and put the object to be weighed in the left cup. Left-handers can switch cups if they like as long as they do so all the time. Using the same cup consistently will minimize weighing error due to instrument bias.

Extension

Hold a class contest. Give a prize to the student that finds a rock weighing closest to 25 grams. (No fair breaking it into pieces.) This will help your students learn to accurately estimate metric weight.

Evaluation

Q: This much graph paper weighs 1 gram. How many squares

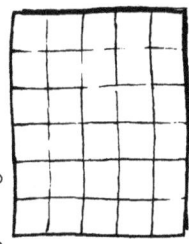

  a. weigh 5 grams?
  b. weigh .1 grams?
  c. weigh .05 grams?

A: a. 150 squares
  b. 3 squares
  c. 1.5 squares

Materials

☐ A soda straw balance.
☐ Gram weights from activity 13.
☐ Supplementary page with reproducible graph paper.
☐ Scissors.
☐ Hand calculator (optional).
☐ Spare change. Either you or your students provide this. One U.S. penny must be older than 1982. If other kinds of coins circulate in your area of the world, modify this worksheet to fit your own needs.

(TO) develop a lighter series of fractional gram weights that complete the paper weight series.

NAME: _____   CLASS: _____   Metric Measuring ( )14

GRAM GRIDS

1 Start with a centered balance and a piece of graph paper. Add just enough graph paper squares to balance your 1 g weight.

1 g

2 Then find how many squares you need to make these other weights.

Write your result here.

FOLLOW THE ARROWS

.2 g = 26.8 squares — MULTIPLY BY 1 → .2 g = 26.8 squares
— MULTIPLY BY 2 → 1 g = 134 squares — DIVIDE BY 10 → .1 g = 13.4 squares
1 g = 67 squares — DIVIDE BY 2 → .5 g = 67 squares — DIVIDE BY 10 → .05 g = 6.7 squares

Your answers may vary according to the paper you use.

3 Cut out the correct number of squares to make the 5 little weights above. (Don't make the 1 g weight because you already have one.)

cut   FOLD   LABEL   .2 g

ALWAYS begin with a CENTERED balance!

4 Check your weights against each other.

☐ Is .1 g heavier than .05 g?
☐ Does .2 g balance .2 g?
☐ Does .5 g balance .2 g + .2 g + .1 g?
☐ Does 1 g balance .5 g + .2 g + .2 g + .1 g?

5 Weigh each coin in grams.

NEW PENNY: 2.5 g
NICKLE: 5.0 g
DIME: 2.3 g
QUARTER: 5.55 g

6 Old U.S. pennies have more copper than new ones. How does this affect their weight?

Older pennies are about .5 g heavier.

OLDER THAN 1982:   NEWER THAN 1982:

7 Predict which weighs more. Show your math in each box.

1 nickle or 2 dimes?
5.0 g   2.3 g
         x2
         4.6 g
5.0 g heavier
☐ correct ☐ wrong

3 dimes or 1 quarter?
2.3 g   5.55 g
x3
6.9 g
heavier
☐ correct ☐ wrong

2 new pennies or 1 quarter?
2.5 g   5.55 g
x2
5.0 g
5.55 g heavier
☐ correct ☐ wrong

2.3 g   5.55 g
5.55 g heavier
☐ correct ☐ wrong

Test each prediction. If you guessed wrong, find your error.

Teaching Notes 15

1. If you know the weight of 1 sugar cube, then allowing for small variations, 2 sugar cubes weigh twice as much; 3 sugar cubes 3 times as much, and so on. You can take advantage of this relationship by weighing only the first sugar cube, then completing the rest of the table mathematically. Or you can directly weigh each group of cubes. Either method produces a good result. Allow your students to follow their own inclinations.

2. This is the first of 4 graphs in this module. Those who are unfamiliar with graphing will benefit from this short review. Mark off a pair of coordinates on your blackboard, then demonstrate how to plot specific data points: first find volume on the horizontal axis and mass on the vertical axis; then plot your point where the perpendiculars from these two locations intersect.

Circle each of your plotted points. When you draw a graph line to show the trend of these points (in this case a straight line), don't draw through the circles. This preserves the clarity of each point so that others can verify you plotted it accurately.

3. Watch out for extensions (extrapolations) that are not properly lined up with the original graph line. This, of course, will introduce error when predicting the mass of 10 cubes.

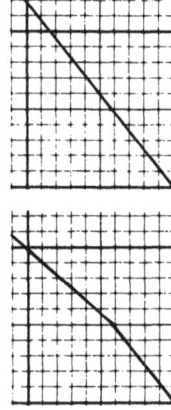

NO   YES

In general, each point (V, M) on the graph line is related by perpendicular lines to two other points—V on the horizontal coordinate and M on the vertical. Knowing any 2 points, you can always find the third. So in this step, given V equal to 10 sugar cubes and (V, M) a point on the graph line, it's easy to find the mass M—about 35.5 grams.

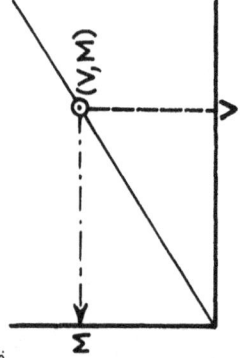

6. Your students can verify the accuracy of their answer. Just read the net weight, in grams on the side of a 2 pound box of sugar cubes. Be sure the cubes are cublet size—252 to a 2 pound box—not the smaller cocktail size.

Discussion

You found the mass of 10 sugar cubes by extending a graph line, by taking a direct proportion and by actually weighing them. These 3 methods form the basis for a good discussion about the reliability of experimental data.

Are all 3 values the same? (Probably not. There are differences due to measuring error. By experimenting carefully you can reduce but not eliminate these errors.) Which method is the most accurate? (Actually weighing the 10 cubes, because it is the most direct. The other two methods rely on masses from a smaller sampling. By multiplying to 10 cubes, you multiply the uncertainty as well.)

Evaluation

Q: This graph shows how the mass of paper clips increases as you add more of them to a balance.

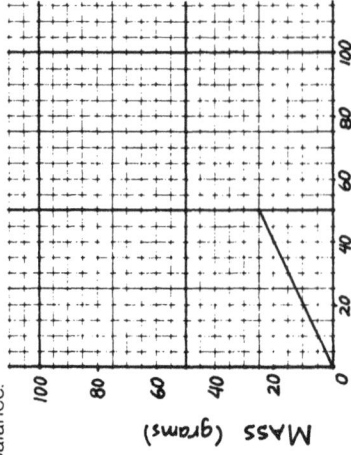

NUMBER OF PAPER CLIPS

a. What is the weight of 30 paper clips?
b. Estimate the weight of 100 paper clips.

A: a. 30 paper clips weigh 15 g.
100 clips weigh 50 g.
b. By extrapolation, 100 clips weigh 50 g.

Materials

☐ A soda straw balance.
☐ A complete set of gram weights from activities 13 and 14.
☐ Sugar cubes.
☐ An index card.
☐ A 2 pound sugar cube box (optional). Your students may like to confirm their answer in step 6 using net weight information on the box.

NAME: _____

CLASS: _____

Metric Measuring (   )15

(TO) use sugar cubes to show how mass changes in direct proportion to volume.
To learn to draw and interpret graphs.

SUGAR CUBE GRAPH

1 First center your balance. Then weigh these sugar cubes and fill in this table.

VOLUME	MASS
0 sugar cubes	0 g
1 sugar cube	3.6 g
2 sugar cubes	7.15 g
3 sugar cubes	10.5 g
4 sugar cubes	14.1 g
5 sugar cubes	17.7 g

2 Plot and circle each point. Draw the best possible straight line through your points using a folded index card.

DON'T DRAW INSIDE YOUR CIRCLES

A folded card makes a good straight edge.

MASS (1 sq. = 1 gram)

VOLUME (5 squares = 1 sugar cube)

3 Use an index card to extend your graph line all the way out. Read the mass of 10 sugar cubes.

GO UP, then OVER

Mass of 10 sugar cubes?

35.5 g

4 Check your prediction using math.

Hint: you know the mass of 5 sugar cubes.

Mass of 10 sugar cubes?

17.7 g
×2
35.4 g

5 Check your prediction on your balance.

Mass of 10 sugar cubes?

35.4 g

10 CUBES   ? GRAMS

6 A box of sugar holds 252 sugar cubes. Find its weight in grams.

$$252 \text{ sugar cubes} \times \frac{35.4 \text{ g}}{10 \text{ sugar cubes}}$$

= 892 grams

(TO) calibrate a styrofoam cup in 20 ml increments. To use this calibrated cup to compare liquid measure to dry measure.

...it assumes vertical planes on all 4 sides of the carton.

See if you can detect this "bulge effect" by listing the results of your whole class on the blackboard. Put liquid measure in one column and dry measure in the other. Throw out volumes that are likely the result of student error. Then find the average of those volumes that are left.

When you examine the numbers, two patterns will likely emerge. First, there will probably be less variation in the dry measure than in the liquid measure. That's because the ruler is more accurate than the calibrated cup. Second, you should notice a consistent excess in the liquid measure amounts, averaging perhaps 50-70 ml more than dry measure. This represents the volume of the bulge.

Demonstration

Do you have access to volumetric lab equipment? If so, assemble an assortment of containers both large and small. Ask your class to guess the volume (in milliliters) of each container. Then demonstrate how to use a graduated cylinder or buret to find the actual volumes. (Don't forget to remeasure the volume of the milk carton as well.)

Find out who is the champion volume estimator for your class: subtract estimated volumes from measured ones to see who maintained the lowest cumulative total.

Evaluation

Q: How many 100 ml cups fill this box? Show your work.

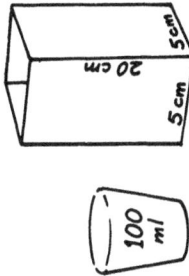

A: First find the volume of the box.

V = (5 cm) (5 cm) (20 cm) = 500 cu cm.

Since 1 cu cm equals 1 ml, the box holds 500 ml, or 5 cups.

Materials

☐ A soda straw balance with its full set of gram weights.
☐ Styrofoam cups.
☐ Water.
☐ A spoon (optional). This enables students to more easily control the amount of water they pour into their weighing cups.
☐ A 12 ounce soda pop can.
☐ A quart milk carton or equivalent.
☐ The 20 cm measuring tape from activity 12.

Teaching Notes 16

---

5. Notice that the pop can and the calibrated measuring cup are illustrated together in this step. This suggests that you can use *both* containers to find the milliliter volume of the milk carton. The pop can has two advantages over the cup. It is more accurate and it has a greater volume capacity.

Insist on recorded data to support each answer. Students can use the back of their worksheets or scratch paper if they need extra space. Here are two different ways to find the volume of the milk carton.

2 cans {	355 ml
	355 ml
3 cans {	355 ml
	355 ml
	355 ml
	355 ml
	1065 ml
	– 80 ml *
	985 ml

2 cups {	100 ml
	100 ml
	+ 80 ml *
	990 ml

*20 ml remains in the last cup.

*Excess left in third can.

5-6. Have your students remembered to record units with each answer? Liquid volume is measured in milliliters, while dry volume is measured in cubic centimeters.

Even though 1 ml equals 1 cu cm, you can expect to record a significant variation between these two measures. This is partly caused by measuring error, of course. The three trial pop can volumes in step 4 establish that the calibrated measuring cup is not very accurate. But there is one other important consideration as well.

When you fill the milk carton with water, notice how it bulges outward on all 4 sides.

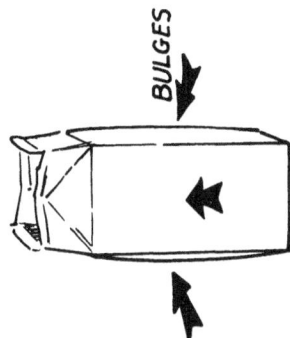

BULGES

Liquid measure fills up this bulge, so it takes this extra volume into account. Dry measure does

---

Metric Measuring ( )16

NAME: _____ CLASS: _____

LIQUID MEASURE

1 Measure a 20 ml volume of water on your balance.

HOW HEAVY IS 20 ml?

How can you measure *volume* on a balance? *When you measure the mass of water in grams, you measure its volume in ml: 1 g water = 1 ml.*

2 Pour this 20 ml of water *inside* another styrofoam cup. Mark the cup to show how high the water reaches.

20 ml:

MAKE A SHORT PENCIL MARK.

POUR INTO ANOTHER CUP

MARK WATER LEVEL

3 Add more water, 20 ml at a time, until you reach 100 ml.

MARK THE NEW WATER LEVEL EACH TIME.

4 Use your metric cup to find how many milliliters fill a pop can.

FIZZ POP

| | Trial 1: | Trial 2: | Trial 3: |
| | 360 ml | 365 ml | 360 ml |

How close did you come to the actual volume printed on the can?

Within 5-10 ml.

5 Find the number of ml in a quart milk carton. Stop adding water at the top fold.

STOP HERE

MILK ONE QUART

990 ml

One ml of water measures one cu cm. So these two volumes should be about the same.

Why don't your LIQUID MEASURE and DRY MEASURE quite agree?

Besides measuring error there is another reason. When you fill the carton, it bulges outward around the middle. You don't get this extra volume with dry measure.

6 Find the volume of this same milk carton in cubic centimeters.

MILK ONE QUART

MEASURE IT!

DRY MEASURE

h = 19 cm
w = 7 cm
l = 7 cm
V = 930 cu cm

Find the difference.

990 ml
– 930 ml
60 ml

TOPS LEARNING SYSTEMS

Teaching Notes 17

1. In this step you'll use your not-so-accurate milliliter measuring cup to accurately determine the density of granulated sugar. How is this possible? By rigging the experiment!

We have preselected two specific masses of granulated sugar—18.2 grams and 36.4 grams—that occupy precisely 20 ml and 40 ml volumes. So when your students pour these amounts into their cups, even though they are poorly calibrated, the sugar will reach somewhere near the 20 ml and 40 ml marks; at least close enough so students will round off to these two values. In this case, rounding off yields precisely the correct answer.

2. Plot each ordered pair as before. Find the volume on the horizontal coordinate and the mass on the vertical coordinate. Follow imaginary perpendicular lines straight out from these positions, then plot your point where they intersect.

This graph establishes the basis for generating a second data table in step 3, so we have added a teacher's check to insure that everyone draws it accurately. Check to see that all three graph points are in the right place. Is the graph line perfectly straight? Watch out for those who idealize the slope at 45° so things will look even.

3. For each 10 ml volume, go straight up until you meet the graph line. Then go straight over to read the mass.

4. In this step your students divide mass by volume to find density. By the time they do this 5 times, we hope the lesson is clear: density doesn't change. Granulated sugar (or any other solid) has only one characteristic density.

Why is this the case? As the volume increases, mass also increases in direct proportion. So the ratio of mass to volume, the density, always divides out to the same number. In graphical terms, density is simply the slope of the graph line. It is constant because the graph line is straight.

Evaluation

Q: One cup of baking soda has a density of 1.1 g/ml. What is the density of 2 cups of baking soda? Explain.

A: The density remains 1.1 g/ml no matter how much baking soda you examine. Whether you have a little or a lot, each ml always weighs 1.1 grams.

Materials

☐ A soda straw balance with gram weights.
☐ Granulated sugar.
☐ A spoon (optional).
☐ A milliliter measuring cup from activity 16.
☐ An index card or straight edge.
☐ A hand calculator (optional).

MAKE-BELIEVE DENSITY

Here is a make-believe demonstration that helps clarify the concept of density. Hold up a styrofoam cup. Announce to your class that it contains imaginary baby oil. Ask a volunteer to smell its sweet aroma. Let another rub some between the fingers. This will create the necessary mind set and provoke curiosity.

Fill a metric cube (from activity 10) until it is brim full of imaginary baby oil. Be careful not to spill any. Then weigh it on an imaginary scale. A folded index card serves as a good prop.

Write your results on the blackboard: *1 ml of baby oil weighs .8 g.* Discuss all the equivalent ways you can say this, thereby introducing the concept of density.

1 ml baby oil = .8 g
Density of baby oil = .8 g in a ml
Density (baby oil) = .8 g per ml
D (b.o.) = .8 g/ml

Repeat this demonstration with cups full of other imaginary substances. Here are some interesting possibilities to consider.

NAME: _____ CLASS: _____

Metric Measuring ( )₁₇

GRANULATED DENSITY

(TO) find the density of granulated sugar. To appreciate that density remains constant — independent of volume changes.

1 To complete this table …

Mass (g)	0 g	18.2 g	36.4 g
Volume (ml)			

… First weigh each mass of *granulated sugar* …

18.2 g

… then pour it into your milliliter cup to find its volume.

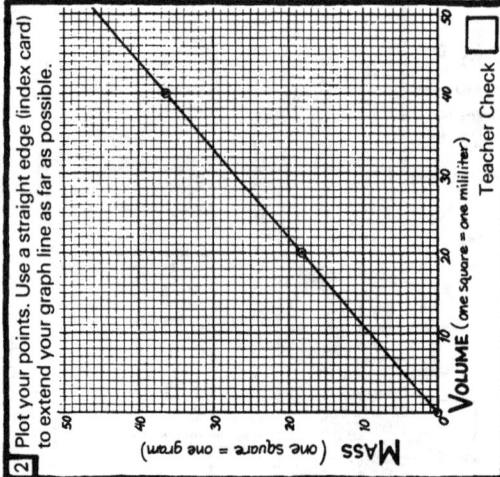

2 Plot your points. Use a straight edge (index card) to extend your graph line as far as possible.

MASS (one square = one gram) / VOLUME (one square = one milliliter)

Teacher Check ☐

3 Read your graph to fill in this column.

GO UP, THEN OVER:

V VOLUME (in ml)	M MASS (in g)	M/V or V√M DENSITY (in g/ml)
10 ml	9.2 g	.92 g/ml
20 ml	18.2 g	.91 g/ml
30 ml	27.4 g	.91 g/ml
40 ml	36.4 g	.91 g/ml
50 ml	46.0 g	.92 g/ml

4 Divide each volume into each mass.

write g/ml (grams in a milliliter) after each number.

5 Is a truckful of sugar more dense than a spoonful? Explain.

No. The table in step 3 above shows that density is constant no matter how much you measure — be it a truckful or a spoonful.

TOPS LEARNING SYSTEMS

Density values (top right)

Density Salt Water: D = 1.4 g/ml
Density Mercury: D = 13.6 g/ml
Density Air: D = .0012 g/ml

salt water · mercury · air

Weigh 1 ml of any substance on your imaginary scale, then announce its gram weight. Let your class practice writing equivalent density equations on scratch paper.

End your discussion with an imaginary pitcher of water and an imaginary barrel of water. Ask which container of water is heavier? The barrel of course.

Now find the density of the *pitcher* of water: each ml weighs 1 gram, so it must be 1 g/ml. Next find the density of the *barrel* of water: each ml still weighs 1 gram, so again it's 1 g/ml. Water has only 1 characteristic density, no matter how much you measure. Even if you sample 100 ml, the density remains unchanged: 100 ml of water weighs 100 grams so the density must be 100 g/100 ml, or 1 g/ml.

The calibrated styrofoam cup has served a useful purpose. But you now need a more accurate volumetric instrument. A laboratory pipet serves as our model and inspiration. Its analog in the world of simple everyday materials is the soda straw.

Your purpose in this activity is to determine the capacity of a soda straw in milliliters. Once you know with accuracy how much a soda straw will hold, you can use it to take accurate volume samples and calculate metric density.

1. Insist on an intelligent estimate, not a wild guess. Ask how your students arrived at their answer. One way to visualize this is to try to imagine how many straw segments fit inside the metric water cube.

Perhaps 4 fit inside, with another one to fill up the spaces in between. So about 5 cm of straw is equivalent to 1 metric water cube. That means 4 cubes would fill a 20 cm straw.

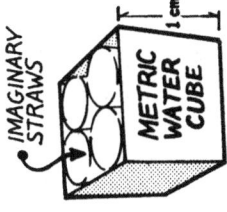
IMAGINARY STRAWS — METRIC WATER CUBE — 1 cm

2. You must provide a deep water container for this activity. If it is too shallow, you cannot easily expel all the air from the straw. If air gets trapped inside, the volume determination will be too small.

It is possible to transfer the water by holding your thumb over just one end. But in doing so, you run the risk of losing small portions of water out the bottom as you transfer it to your balance. By closing both ends as you transfer the water more securely inside.

3. Some may weigh consecutive strawfuls, first one, then two combined, then three combined. This is incorrect. The idea here is to weigh a single strawful of water in 3 separate trials, then to select the mass you think comes closest to the true value.

Evaluation

Q: Here is an unopened can of soda pop. Estimate its total weight in grams. Explain your reasoning.

POP — 355 ml

A: The soda pop is mostly water. So it's 355 milliliter volume weighs about 355 grams. The empty pop can weighs about 5 or 6 pennies, perhaps 20 grams. Together, the can and liquid weigh about 375 grams.

Materials

□ Plastic soda straws. We recommend Glad brand (7¾ inch by 15/64 inch dia.) manufactured by Union Carbide, if available. These have volumes very close to 5 ml, a nice round whole number to work with.
□ A metric water cube from activity 10.
□ A sink or tub with water. It should be wide enough to accomodate at least one pair of hands or more. It must be deep enough to keep the straw entirely submerged in a more or less vertical position (at least steeper than 45°). See teaching notes 2 above.
□ A styrofoam cup.
□ A soda straw balance with gram weights.

Teaching Notes 18

Student Worksheet (left side)

(TO) find the volume of a soda straw in milliliters so it can be used as a volumetric tool.

NAME: _____

CLASS: _____

Metric Measuring ( )18

METRIC STRAW

1 How many milliliter cubes of water do you think will fill a straw? Use your metric water cube to make an educated guess.

Your Guess: ___ ?

varied answers

2 Practice pouring straws full of water into a cup. Make sure the straw is completely full and that no water drips into the cup from your hands.

Hold under water at a slant so *all* the air escapes. — Cover both ends with your fingers *before* you lift it from the water. — Take your fingers off *both* ends to empty your strawful.

Keep practicing until you get good at it! □ TEACHER CHECK

3 Center your balance, then weigh a strawful of water. Repeat at least 2 more times to see if you get close to the same value each time.

CENTER your balance EACH TIME before you weigh.

TRIAL:
1 one strawful = 4.9 grams
2 one strawful = 5.0 g
3 one strawful = 5.1 g

Decide what value to use as your

Official Mass: 5.0 grams

4 How many *milliliter* of water does your straw hold?

Remember: 1 ml water = 1 g

Official Volume: 5.0 ml

How close was your guess in step 1?

varied answers

☞ SAVE YOUR SODA STRAW

Teaching Notes 19

6. Directions for this step are intentionally vague. You can, if you wish, copy the table in step 4, then fill it in with values for corn meal. Or you can follow a more independent strategy.

Simply make one density determination and plot that point on the graph. You already know that zero ml of corn meal weighs zero grams. Since two points determine a straight line you can draw in the graph line.

Whatever strategy your students follow, insist they write units. In doing so they are less likely to confuse the various units of measure.

Extension

You already found the density of sugar using your calibrated styrofoam cup. Find it again with your soda straw. See if you get close to the same answer.

Evaluation

Q: Here is volume and mass data for a certain brand of baby oil. (Supply graph grid like the one below.)

Volume	5 ml	10 ml	15 ml	20 ml
Mass	4 g	8 g	12 g	16 g

a. Plot this data and draw a graph line.
b. Find the density of this baby oil.

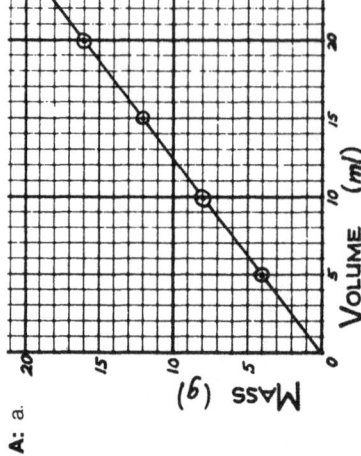

A: a.

MASS (g) vs VOLUME (ml)

b. Divide any mass by its associated volume. The density is always .8 g/ml.

$$\frac{4\ g}{5\ ml} = \frac{8\ g}{10\ ml} \cdots = .8\ g/ml.$$

Materials

☐ A plastic soda straw with a known milliliter volume from activity 18.
☐ Cellophane tape.
☐ Table salt.
☐ A styrofoam cup.
☐ A soda straw balance with gram weights.
☐ An index card or straight edge..
☐ Cornmeal.
☐ Hand calculator (optional).

Knowing the volume of your plastic soda straw, you can now sample strawsful of material to the nearest tenth of a milliliter. Then you can weigh it on your balance to the nearest tenth of a gram. You have the wherewithal, then, to make accurate density determinations in grams per milliliter.

1. Since you only use the straw to measure dry granulated solids, you can tightly seal off one end with just an overlapping piece of tape. The straw may flatten slightly when you squeeze this tape together. But this will have a negligable effect on its overall volume. Treat the straw, then, as if it were a long narrow test tube.

If humidity is a problem in your area, causing table salt to lump and cake, be sure to store it in a dry, sealed container. Add a few soda crackers to absorb moisture. This ensures that the salt will pour freely and rest compactly within the narrow confines of the straw.

2. Fill the straw with salt as if you were taking a core sample from the earth with a hollow pipe. Keep it tilted toward the horizontal so that once inside, the salt doesn't slide back out. Tap the sealed end on the table occasionally to dislodge any air pockets and pack the salt firmly inside.

Continue filling and tapping it down until the salt reaches somewhere near the top of the straw. You'll have to top off the remaining centimeter or so by adding salt pinches with your fingers.

4. You can weigh just one strawful of salt, then fill in this entire table. Two strawsful weigh twice as much; three strawsful 3 times as much and so on. The density remains constant no matter how many strawsful you measure.

In practice most students will probably be too involved in the measuring, weighing and calculating to think about taking shortcuts. That's fine. They need to experience the process of analytical science more fully—to feel the sense of accomplishment associated with finding all those masses, then dividing to find all those densities.

If you decide to provide calculators, don't tolerate a long string of meaningless numbers in the 8-digit display. Round off each answer to the nearest hundredth.

There is only 1 density for table salt. The response box underneath the table requires students to examine all their various experimental values, then decide on one official value. They might select the mode, the answer that appears most often. Or they can compute an average. If a density value seems to fall far outside the norm, it should be thrown out before taking an average.

NAME:

CLASS:

Metric Measuring (   )₁₉

(TO) calculate the density of table salt and corn meal, then express each result on a graph.

THE LAST STRAW

1 Wrap tape around a clean dry straw so it hangs over the end. Squash the tape flat to seal the end.

OVERLAP TAPE

2 Fill the straw to the top with salt.

Dig it into some salt...

... then tap it down.

CLOSED END

3 Weigh the salt on your centered balance.

Be sure to shake out all the salt.

4 Fill in this table for salt. (You already found the straw's volume.)

Number of Strawsful	VOLUME (in ml)	MASS (in g)	$\frac{M}{V}$ or V/M DENSITY (in g/ml)
1	5 ml	6.9 g	1.38 g/ml
2	10 ml	13.6 g	1.36 g/ml
3	15 ml	20.6 g	1.37 g/ml
4	20 ml	27.1 g	1.36 g/ml

Decide what value to use as your . . . . Official Density 1.37 g/ml

5 Graph your results.

MASS (one square = one g) vs VOLUME (one square = one ml)

SALT — CORN MEAL

6 Find the density for corn meal. Plot your result on this same graph and label each line.

For your data:

5 ml corn meal = 3.9 g
20 ml corn meal = (4) (3.9 g)
= 15.6 g

$$D = \frac{3.9\ g}{5\ ml} = .78\ g/ml$$

Official Density .78 g/ml

(TO) correlate the density of a substance with its floating and sinking characteristics.

NAME:

CLASS:

Metric Measuring ( )20

SINK OR SWIM?

1 You already know these densities. Fill them in:

2 Complete each little table.

Use your head, not your balance

WATER: g/ml		SUGAR: g/ml		SALT: g/ml		CORN MEAL: g/ml	
VOLUME	MASS	VOLUME	MASS	VOLUME	MASS	VOLUME	MASS
0 ml		0 ml		0 ml		0 ml	
20 ml		20 ml		20 ml		20 ml	

3 Draw and label a graph line for each item.

MASS (one square = one g)

VOLUME (one square = one ml)

4 Every material has its own density. What does the slope (steepness of the line) tell you about density?

which way does density increase?

MASS
VOLUME

5 Notice that the density line for water separates salt on one side from sugar and corn meal on the other. Predict what floats in water and what sinks.

6 Test your prediction for sugar. Wrap a cube completely in tape so no corner is left uncovered.

If you see air bubbles come out in water, you have a leak!

Does it float or sink as you predicted?

TOPS LEARNING SYSTEMS

Extension

Test the floating and sinking characteristics of salt and corn meal. Place small amounts of each in plastic lunch bags, then twist closed and tie each with string. Cut off as much excess plastic and string as possible.

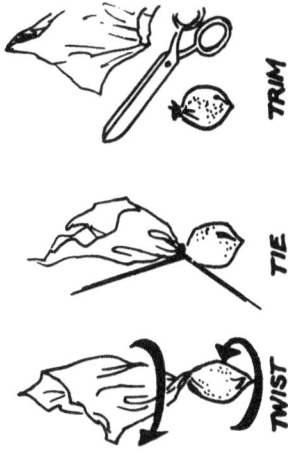

TWIST   TIE   TRIM

If you seal these tightly with no air spaces inside, the corn meal will float on water and the salt will sink, just as the density data suggests.

Evaluation

Q: The density line for water is the one labeled b. Which density line represents styrofoam? Explain how you know.

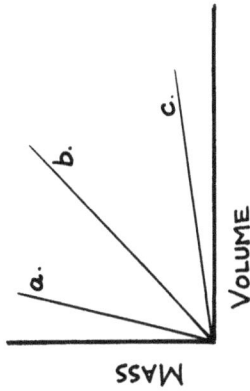

VOLUME

MASS

a.   b.   c.

A: Line c. Because styrofoam floats on water, it is less dense than water. So its graph line has less slope.

Materials

☐ A metric water cube from activity 10. This may help students understand that the density of water equals 1 g/ml.
☐ Activity sheets 17 and 19. Students need to refer to their experimental density values for sugar, salt and cornmeal.
☐ An index card or straight edge.
☐ Sugar cubes.
☐ Cellophane tape.
☐ A styrofoam cup with water.

Teaching Notes 20

---

1. The density of water is easy to find. Just examine your metric water cube. One ml of water weighs 1 gram. So the density must equal 1 g/ml.

You found the density of sugar in activity 17. The density of salt and corn meal come from activity 19.

2. Since nothing weighs nothing, each zero volume has zero mass as well. Finding the mass for 20 ml is almost as easy. Since you know the mass for 1 ml, all you have to do is multiply by 20.

In general, when you multiply density by volume you always get mass:

$$\left(\frac{gm}{ml}\right)\left(\frac{ml}{1}\right) = \left(\frac{gm}{1}\right)$$

(Density) (Volume) = Mass

3-4. Two points determine a straight line. So each little table contains enough information to plot the entire graph line. One of these points (0,0) is common to all 4 lines. The other 4 ordered pairs are all located on the vertical line that defines 20 ml.

By labeling each graph line, it's easy to correlate an increase in density with an increase in slope. Salt has the steepest slope and the highest density. Corn meal has the shallowest slope and the lowest density.

5. A graph is *not* a glass of water! Watch out for those who think it is. The illogic goes something like this: On the graph, salt is above water while sugar and corn meal are below. So salt floats while sugar and corn meal sink. Wrong!

6. Anyone who drops a sugar cube into a cup of coffee knows that it sinks like a stone. But that's only because it rapidly absorbs water. Wrap the cube completely in tape, so no water can get inside, and the sugar will float.

To wrap it airtight, you'll need to cover all 8 corners and all 12 edges with tape. Miss even one and the cube will go under, leaving a trail of tiny bubbles as it sinks. Students who fail to make it float the first time will want to try it again.

HOW TO REMOVE WORKSHEETS

Perforated worksheets don't always work like they are supposed to.

This book is designed not only to make your science lessons run smoothly, but to make the worksheets pull out smoothly as well. Our pages are "perfect bound" in the same manner as single sheets of stationary are attached to a writing pad. You can remove worksheets from this book just like pulling sheets off a pad — well, almost.

We didn't want our book to shed leaves like a tree. So we ordered the perfect binding very strong. To remove the worksheets cleanly and quickly, be sure to follow one of these two special procedures.

One-at-a-time:

Start from the *back* of the book. *Pull* the top sheet off as illustrated. Don't tear. Proceed to the next until you remove all the worksheets.

The top sheet will probably be glued most securely to the binding. Sheets underneath should pull off more easily. Do not attempt to remove pages from the middle of the book *first*. This is often difficult to do (even on a scratch pad).

PULL GENTLY OUTWARD— LAST PAGE FIRST. DON'T TEAR!

Radical Surgery:

Place a sharp knife on this very page with the edge facing the binding. Close the book, and pull the knife through the binding to cleanly remove all the worksheets. Strip off the back cover and separate each page.

REPRODUCIBLE
STUDENT
ACTIVITY SHEETS

METRIC STAIRS (1)

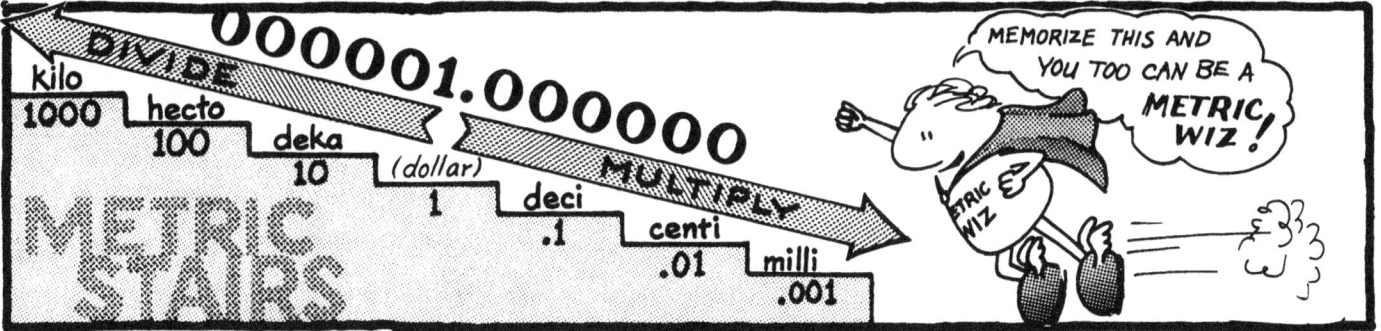

DIVIDE 000001.000000 **MULTIPLY**

kilo 1000
hecto 100
deka 10
(dollar) 1
deci .1
centi .01
milli .001

METRIC STAIRS

MEMORIZE THIS AND YOU TOO CAN BE A METRIC WIZ!

1 On a piece of lined notebook paper, write 6 equations, 1 for each metric step.

WRITE THIS TOP STEP FIRST

kilo = 1000

2 Cut out this cube pattern. Fold and tape it around a SUGAR CUBE!

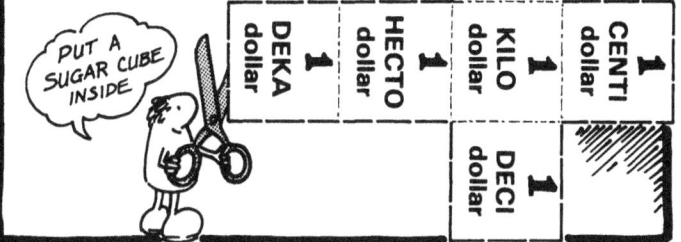

PUT A SUGAR CUBE INSIDE

	1 MILLI dollar	
1 DEKA dollar	1 HECTO dollar	1 KILO dollar	1 CENTI dollar
	1 DECI dollar	

3 Roll 6 *different* cash amounts. Write how much each is, just like the example.

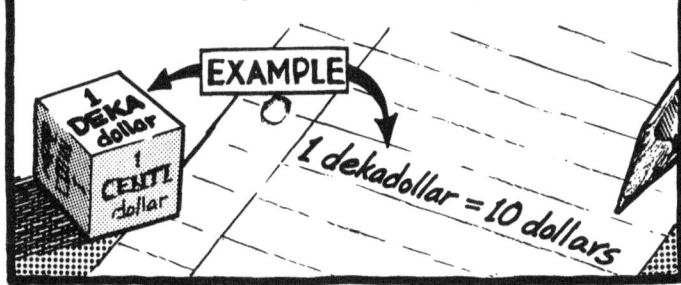

EXAMPLE

1 DEKA dollar / 1 CENTI dollar

1 dekadollar = 10 dollars

4 Cut out this cube pattern with boxes. Fold and tape it around a sugar cube like you did with the first.

	DECI dollars ⏸	
CENTI dollars ⏸	KILO dollars ⏸	HECTO dollars ⏸	DEKA dollars ⏸
	MILLI dollars ⏸	

(BOXES)

5 Throw this "box" cube 8 times. Each time it lands find the missing number that *makes one dollar*. (STUDY THE EXAMPLE!)

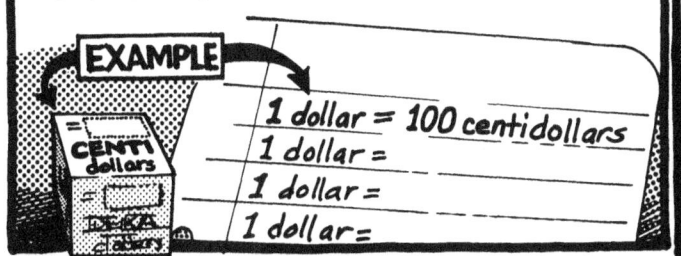

EXAMPLE

CENTI dollars / DEKA dollars

1 dollar = 100 centidollars
1 dollar =
1 dollar =
1 dollar =

6 Toss *both* cubes 12 times. Copy each equation (first the "1" cube, then the "box" cube), then find the missing number.

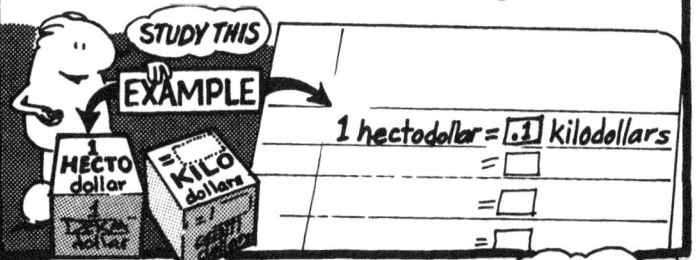

STUDY THIS EXAMPLE

1 HECTO dollar / 1 KILO dollars

1 hectodollar = .1 kilodollars

7 Two can play metric dice: roll one pair of dice; write down answers; compare.

If answers agree, move a paper clip forward 1 space.
If answers disagree, move back 2 spaces.

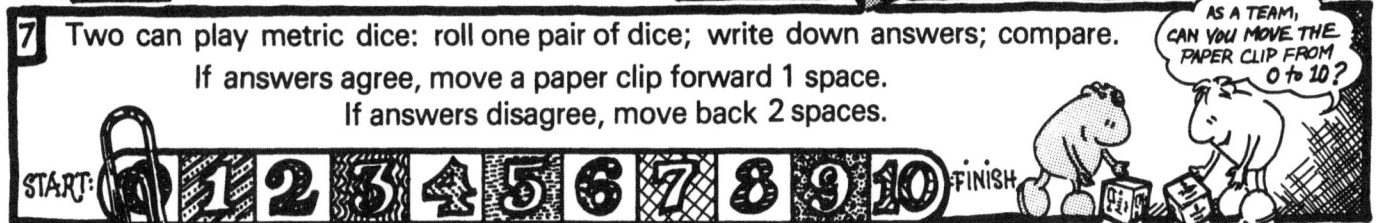

AS A TEAM, CAN YOU MOVE THE PAPER CLIP FROM 0 to 10?

START: 0 1 2 3 4 5 6 7 8 9 10 FINISH

TOPS LEARNING SYSTEMS

SWEET LENGTHS

Ruler scale (left side):
0, 1, 2, 3, 4, 5, 6, 7, 8, 9, 10

100 centiLengths
1 centiLength
10 centiLengths = 1 Length

SWEETRULER →

1 Cut out this ruler.

SWEETRULER

How many sugar lumps fit between 0 and 10 on this ruler?

2 Tape this ruler to the *lower left* corner of a long strip of paper.

PLACE IT ON THE EDGE.

3 Use a row of 10 sugar lumps to number a full "sweetruler" with 100 lumps.

MAKE A PENCIL MARK BETWEEN EACH LUMP AT THE EDGE OF THE PAPER.

to 100

4 Cut your sweetruler off at the 100 mark. Write "1 LENGTH" on your ruler in large bold letters.

I HEREBY NAME THEE ONE LENGTH

1 LENGTH

5 Your ruler is 1 Length. Is your room longer than a *deka* Length?

10 of these is 1 decalength.

ONE LENGTH

6 Your ruler is 1 Length: Name the distance equal to 100 sweetrulers laid end to end.

? Lengths

7 Your ruler is 1 Length: Name the distance equal to 1000 sweetrulers laid end to end.

1000 LENGTHS

8 Your ruler is 1 Length:

Name the distance equal to 0.1 sweetrulers.

How many sugar lumps measure this long?

9 Your ruler is 1 Length:

Name the distance equal to 0.01 sweetrulers.

How many sugar lumps measure this long?

10 Your ruler is 1 Length:

Name the distance equal to 0.001 sweetrulers.

How many milliLengths are in your sweetruler?

WRITE YOUR NAME ON YOUR SWEETRULER AND SAVE IT!

TOPS LEARNING SYSTEMS

NAME: CLASS:

ONE DIMENSION

1   Use your sweetruler to measure each distance.

Answer in **CENTI**LENGTHS (cL)

Write your answer, PLUS ITS UNIT, in each box.

Answer in **MILLI**LENGTHS (mL)

a.

b.

c.

d.

e.

f.

g.

h.

Answer in **LENGTHS** (L)

Answer in **DECI**LENGTHS (dL)

i.   The length of your room?

j.   The length of your desk?

k.   The height of the doorway?

2   This map shows the path Terry takes each day to school. Let the side of a sugar cube (1 centiLength) stand for a hectoLength . . .

SCALE: 1 centiLength (sugar) equals 1 hectoLength (on map)

SCHOOL   LAKE   HOME

Answer in **HECTO**LENGTHS (hL)

Answer in **KILO**LENGTHS (kL)

a.   How far must Terry walk to school?

b.   How long is the lake?

c.   How far would a crow fly from school to Terry's house?

TOPS LEARNING SYSTEMS

SQUARED DIMENSIONS

1. Measure both sides of this box, then multiply to find the area.

Write measurements AND THEIR UNITS in the table below.

A

2. Measure this box and all the others in the same way.

B

Don't forget the UNITS!

C

D

E

To find AREA, just multiply.

$$(\text{length}) \times (\text{width}) = \text{Area}$$
$$\text{cL} \times \text{cL} = \text{sq. cL}$$

Double check your math answers — cover each area with sugar lumps and count the squares!

WRITE THE UNITS

box	length × width =		by multiplying	by covering with sugar lumps
A	×	=		
B				
C				
D				
E				

Area

Did you write all the UNITS?

TOPS LEARNING SYSTEMS

CUBED DIMENSIONS

1. Imagine that this box is covered with sugar cubes stacked **1 High**.

Measure all 3 sides and find the volume.

Write measurements in the table below...

2. This box is stacked **2 High**, and the others 3 high, 4 high 5 high and 1 high.

Find each measure and complete the table.

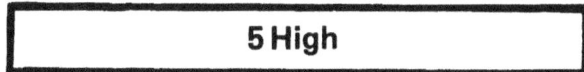

3 High

1 High

4 High

5 High

Multiply to get AREA, then multiply again to get VOLUME:

(Area) x *(height)* = *Volume*

cL x cL x cL = cu. cL

Double check your answers — build each volume and count the cubes.

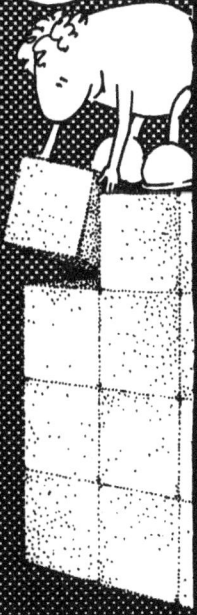

WRITE THE UNITS

box	length	×	width	×	height	=	Volume by multiplying	by counting cubes
1		×		×		=		
2								
3								
4								
5								
1								

TOPS LEARNING SYSTEMS

NAME: CLASS:

LENGTH...AREA...VOLUME

1 Fold two 3 x 5 index cards exactly in half the short way.

FOLD

2 Tape the cards together to form a box.

3 Measure this box to the nearest centiLength with sugar cubes. Be sure to *include the correct unit* with each answer.

a. height = ...

b. length = ...

c. width = ...

d. area of side = ...

e. volume = ...

f. area of top = ...

g. length of all 12 edges = ...

h. surface area = ...
(all 4 sides)

Height!   Volume!   Area!   Length!   Width!

4 How many sugar squares fit on the surface of your desk? Show your math.

5 How many sugar cubes could a grocery bag hold? Show your math.

TOPS LEARNING SYSTEMS

BUILD A STRAW BALANCE

1 Double some string to make a long U-shape.

Cut its length equal to 1 straw plus 1 paper clip.

2 Tie the string in a loop.

KEEP THESE ENDS SHORT.

3 Lay your straw on the one shown at right so each end is the exact same distance from the same letter . . .

SAME LETTER

IHGFEDCBA

ABCDEFG

MARK CENTER

. . . Lightly poke the exact center and show it to your teacher.

☐ Teacher Check

4 Push your string loop through your straw so it hangs out both ends.

Keep the knot at one end.

KEEP THE KNOT HANGING OUT.

5 Cut open a new straw along its full length . . .

. . . Then slide it inside your first so the string is pressed *between* both straws.

HOLD ONTO THE STRING— DON'T LET IT SLIP INSIDE.

CENTER

6 Push a pin *straight* through the middle of the straw where you poked it before. Show it to your teacher.

PUSH STRAIGHT THROUGH ← → NOT CROOKED

☐ Teacher Check

7 Fold tape over the ends of a clothespin. Each piece should stick out past the end about as wide as a paper clip.

PINCH TAPE FLAT

8 Cut out a narrow strip from the center of the tape.

LOOKS LIKE EARS!

CUT TO THE WOOD

9 Clamp the clothespin to the pull-tab on a pop can.

REST PIN IN SLOTS

BE **SURE** THE LOOPS HANG OUT THE **BOTTOM**...

...TURN THE STRAW OVER IF THEY DON'T.

TOPS LEARNING SYSTEMS

STRAW BALANCE, CONTINUED

10 Write your name on this tag: [   ] Cut it out and tape it to the *top* of your beam.

ALWAYS KEEP YOUR NAME ON *TOP*, LOOPS AT THE *BOTTOM*.

11 Bend out the arms of two paper clips just a little. Hang them on each end to form hooks.

12 Cut 2 strings about as wide as this paper. Tie each one around your pencil in a knot, then slide it back off.

TIE

SLIP OFF

13 Stretch each string across the mouth of a styrofoam cup and tape it near the rim. Keep the loop in the center, and let each end hang loose.

CENTER

LOOSE ENDS

14 Pull up the loose ends and tape them a second time. Trim off the excess.

TAPE THEN *TRIM*

15 Hang each cup on the balance from its little loop. Add tape to the higher cup until the beam balances exactly level.

THAT'S ENOUGH TAPE!

16 Write "knot" on the cup that hangs from the knotted end of the string. Always hang this "knot cup" on the knot side of the balance; never switch the cups.

KNOT

Knot

17 Which is heavier, a sugar cube or a penny? First guess, then test.

Guess

Test

18 Roll up some tape so that just one end is sticky.

LEAVE A STICKY TAB

FOLD A "HANDLE"

19 Empty your balance. Center it back to level by resting the tape tab somewhere on the high side of the beam.

TAPE TAB — Move to where the beam levels

20 How many paper clips balance a penny?

★ **SAVE YOUR BALANCE** ★

TOPS LEARNING SYSTEMS

SWEET MASSES

1 Let's say a sugar cube weighs one "MASS".

I hereby name thee **ONE MASS !**

2 If a sugar cube weighs 1 Mass, how many cubes weigh a kiloMass?

Complete the table.

PREFIX:	MEANING:	UNIT:	HOW MANY SUGAR CUBES:
kilo	thousand	kiloMass	
hecto			
deca			
—	—	Mass	1 sugar cube
deci			
centi			
milli			

3 Answer each question using your balance. Remember to . . .

. . . keep your name tag UP and the strings DOWN . . .

. . . center your empty balance with the tape tab each time before you weigh.

a. How many paper clips weigh 1 Mass?

= one Mass

b. How many pennies weigh 1 dekaMass?

= one deka-Mass

c. How many rice grains balance a paper clip?

= one

d. You know how many rice grains balance a paper clip and how many paper clips weigh 1 Mass. Calculate how many rice grains weigh 1 Mass.

e. Is a rice grain larger or smaller than a milli Mass? Explain how you know.

TOPS LEARNING SYSTEMS

SWEET VOLUMES

1 Cut 4 squares from lined notebook paper.

Make each square 10 centiLengths on each side . . .

2 Tape the 4 squares together to form a "sweetbox".

3 Let's call the space occupied by your sweetbox 1 VOLUME.

Label your box this way.

Sweetbox 1 Volume

METRIC WIZ

4 How many sugar cubes fill your sweet-box?

5 Complete this table.

1 sweetbox	1 Volume		sugar cubes
.1 sweetbox	1 deciVolume		sugar cubes
.01 sweetbox			sugar cubes
.001 sweetbox			sugar cubes

1 Volume = ▢ milli-Volumes

6 Cover this milliVolume box pattern completely with tape to make it waterproof. Then cut it out.

1 mV

OVERLAP THE TAPE STRIPS.

7 Fold the flaps so the tape is on the inside of the box.

Seal all 4 edges with tape.

OVER HERE!

8 Pour 10 cubes of water into a styrofoam cup. Mark the 10 mV level on the inside.

10 mV

9 How many *cups* fill your sweetbox? Show your work.

FIRST find how many mV's are in one FULL cup!

STYROFOAM CUP

10 You can measure volume in 2 ways:

1 USE A RULER . . . . . . this is called DRY MEASURE

2 USE A MEASURING CUP or CUBE . . . . . . this is called LIQUID MEASURE

How many *cubic centiLengths* in . . . a sugar cube? . . . a sweetbox?	How many *milliVolumes* in . . . a sugar cube? . . . a sweetbox?

TOPS LEARNING SYSTEMS

SUGAR CUBES/WATER CUBES

SWEET measure works just like METRIC measure.

Sweet measure is based on a SUGAR cube...

SUGAR CUBE:

WATER CUBE:

├─ Actual Size ─┤

... METRIC measure is based on a WATER cube!

To measure in METRICS we must replace sugar cubes with water cubes, and learn just

3 NEW WORDS

---

1 | Mass is in GRAMS.

a. A Mass equals 1 sugar cube. So how heavy is a **gram?**

b. Cut out this metric water cube pattern. Fold and tape it together using *tiny* pieces of tape.

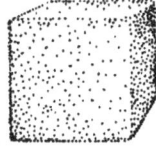

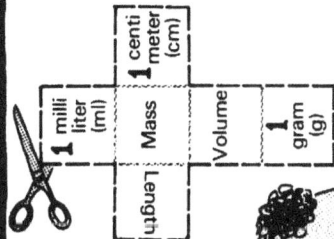

Filled with WATER, this cube weighs 1 GRAM

---

2 | Length is in METERS.

a. A Length (sweetruler) equals 100 sugar cubes. So how long is a **meter?**

b. Use your metric water cube to make a strip of paper 1 meter long. Number each 10 spaces and label it "1 Meter".

Each unit is 1 metric water cube long!

METRIC WATER CUBE

LABEL

1 METER

---

3 | Volume is in LITERS.

a. A Volume (sweetbox) equals 1000 sugar cubes stacked together. So how big is a **liter?**

b. Use your metric water cube and lined notebook paper to make a liter cube. Label it "1 Liter".

1000 WATER CUBES

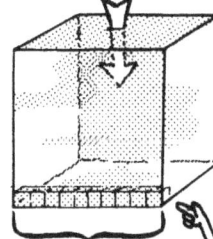

A cube 10 times longer is 1000 times bigger!

10 WATER CUBES

---

4 | Memorize the length, mass and volume units of your metric water cube ...

OPPOSITE SIDES: GO TOGETHER

... Then toss the cube. Read the top to predict the bottom.

Mark ⓪ if you got it right;

🅇 if you got it wrong...

20 TOSSES:

SAVE YOUR 3 METRIC MODELS

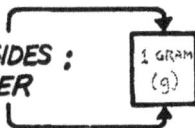 LEARNING SYSTEMS

METRIC STAIRS (2)

1 Write the correct metric unit on each stairstep. Don't abbreviate.

Remember: kilo-, hecto-, deka-, deci-, centi-, milli- ...

1,000 meters ★

1,000 grams ★

1,000 liters

100 meters

100 grams

100 liters

10 meters

10 grams

10 liters

Units with stars are used often. Learn them!

1 meter **meter** ★

1 gram **gram** ★

1 liter **liter** ★

0.1 meter

0.1 gram

0.1 liter

0.01 meter

0.01 gram

0.01 liter

0.001 meter

0.001 gram

0.001 liter

2 Write equal measures in each box below. Use the decimal stairs to help you.

Stand on the first "step" given in each box.
MULTIPLY *as you step* **DOWN** *(move decimal right).*
DIVIDE *as you step* **UP** *(move decimal left).*

1 meter =	**1 kilometer =**	**1 centimeter =**
0.001 km	m	km
cm	cm	m
mm	**1,000,000 mm**	mm

1 gram =	**1 kilogram =**	**1 milligram =**
kg	g	kg
mg	mg	g

1 millimeter =
km
m
cm

1 liter =
ml

1 milliliter =
l

3 On your meter tape, draw and label a line equal to ...

... 1 decimeter;
... 1 centimeter;
... 1 millimeter.

4 On your liter model ...

... Write how many milliliters it holds.

... If it were filled with water, write how much it would weigh.

1 liter

TOPS LEARNING SYSTEMS

DRY MEASURE

1 Use the metric ruler on the side of this paper to cut a string 1 meter long.

Measure the length of your room in meters.

HOW MANY STRING LENGTHS?

2 Cut string to the correct length and tape it to each box.

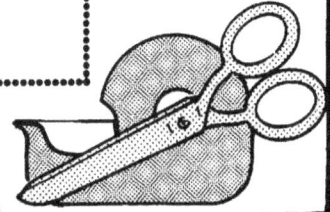

centimeter:

decimeter:

millimeter:

METRIC RULER

3 Now cut out your metric ruler and find each measure on this box. Write numbers *and* units.

length = ...

width = ...

height = ...

area of front = ...

area of side = ...

volume = ...

FRONT   SIDE

width   height   length

HINT: USE YOUR METRIC WATER CUBE TO ANSWER THESE.

How many milliliters of water fill this box? How do you know?

When filled with water, how much does this box weigh? How do you know?

4 Imagine having 11 of these boxes filled evenly to the top with gold dust! Would you trade them for a liter of gold dust?

SHOW YOUR MATH!

SAVE YOUR RULER

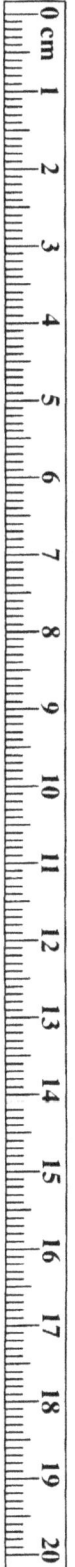

TOPS LEARNING SYSTEMS

GRAM QUARTERS

1 Center your balance, then add 20 grams to your left cup.

(20 g = 8 new U.S. pennies)

2 Fold 5 sheets of paper together the long way. Slip the *folded* side under the *right* paper clip so the cup leans out of the way.

5 SHEETS

FOLD TOGETHER

CUP

3 Is this paper heavier than 20 g?

4 Cut *parallel* strips off the 5 folded papers until they exactly balance 20 g. Cut only *towards* the fold.

FOLD

CUT JUST A LITTLE AT A TIME...

UNTIL IT BALANCES LEVEL.

5 Fold each paper into 4 *equal* quarters. Number them from 1 to 20. How much does each numbered quarter weigh?

6 Fold 10 g of paper (10 quarters) as small as you can . . .   Tape it together and label it.

GATHER 10 QUARTERS

FOLD TIGHTLY

USE VERY LITTLE TAPE

10 g

7 Make these other weights with the rest of your paper quarters.

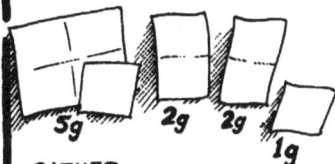

5g   2g   2g   1g

GATHER THESE QUARTERS

Don't tape the small ones

5g   2g   2g

1g

FOLD AND LABEL

8 Make a 20 g weight by pouring salt into a sandwich bag.

Twist the plastic closed and tie it in a knot.

20g Paper Weights

Sandwich Bag

9 Check your weights against each other.

☐ Does 2g balance 2g ?

☐ Does 5g balance 2g+2g+1g?

☐ Does 10g balance 5g+2g+2g+1g?

☐ Does 20g balance 10g+5g+2g+2g+1g?

SAVE YOUR WEIGHTS !

TOPS LEARNING SYSTEMS

GRAM GRIDS

1 Start with a centered balance and a piece of graph paper. Add just enough graph paper squares to balance your 1 g weight.

1g

2 Write your result here.

Then find how many squares you need to make these other weights.

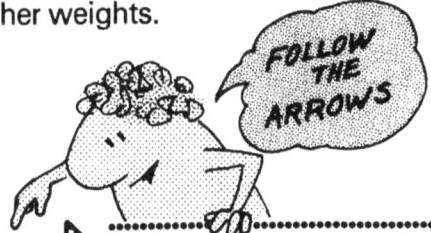

FOLLOW THE ARROWS

.2 g = _____ squares

MULTIPLY BY 1

1 g = _____ squares → DIVIDE BY 2 → .5 g = _____ squares

DIVIDE BY 10

DIVIDE BY 10

.2 g = _____ squares ← MULTIPLY BY 2 ← .1 g = _____ squares

.05 g = _____ squares

3 Cut out the correct number of squares to make the 5 little weights above. (Don't make the 1 g weight because you already have one.)

CUT   LABEL .2 g

FOLD

4 Check your weights against each other.

☐ Is .1 g heavier than .05 g?

☐ Does .2 g balance .2 g?

☐ Does .5 g balance .2 g + .2 g + .1 g?

☐ Does 1. g balance .5 g + .2 g + .2 g + .1 g?

5 Weigh each coin in grams.

ALWAYS begin with a CENTERED balance!

NEW PENNY: ....

NICKLE: ....

DIME: ....

QUARTER: ....

6 Old U.S. pennies have more copper than new ones. How does this affect their weight?

OLDER THAN 1982:   NEWER THAN 1982:

??

7 Predict which weighs *more.* Show your math in each box.

1 nickle or 2 dimes?	3 dimes or 1 quarter?	2 new pennies or 1 quarter?
☐ correct ☐ wrong	☐ correct ☐ wrong	☐ correct ☐ wrong

SAVE YOUR WEIGHTS!

Test each prediction. If you guessed wrong, find your error.

TOPS LEARNING SYSTEMS

Supplement for Worksheet 14:   **GRAM GRIDS**

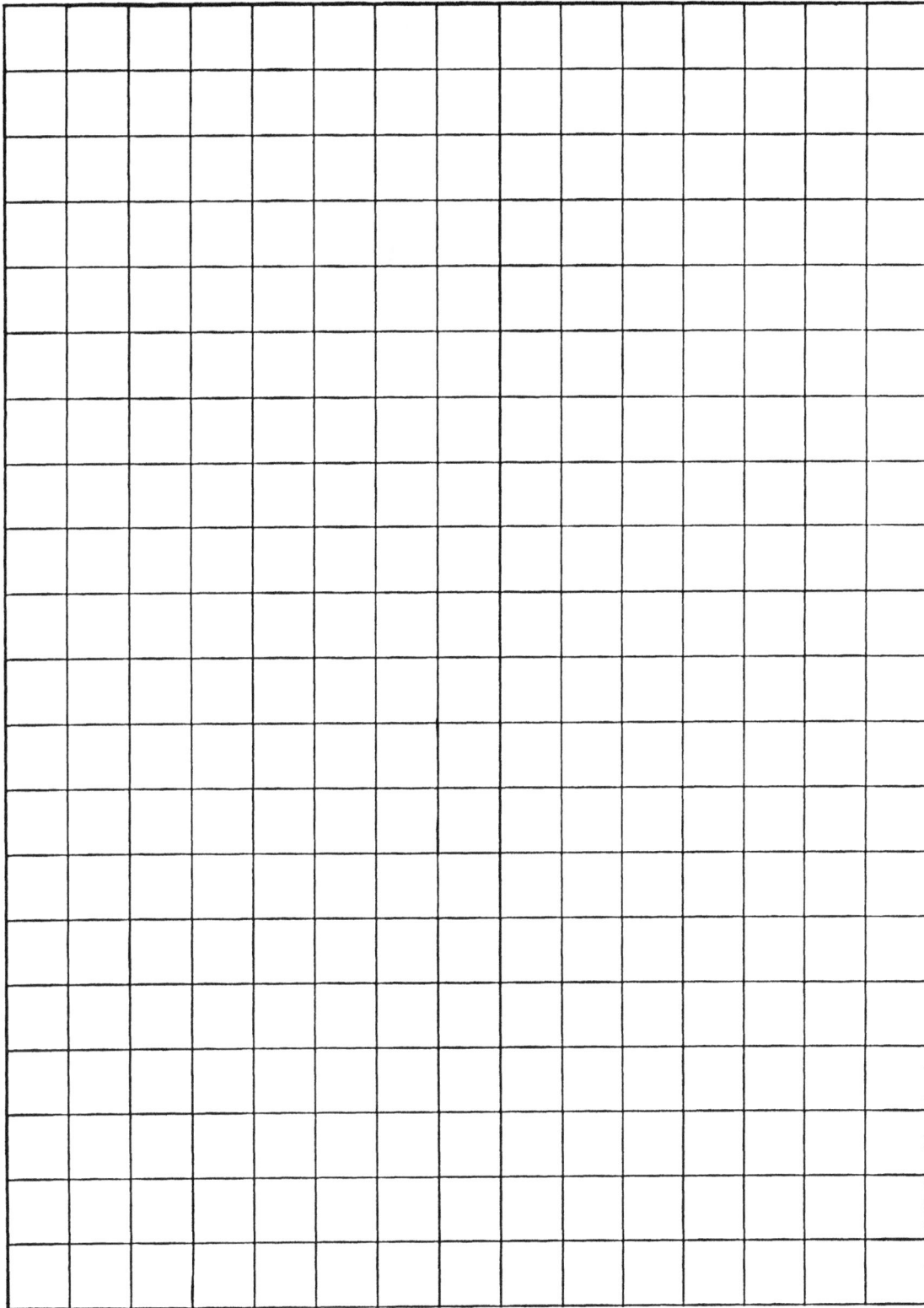

ALTERNATE SWEETRULER

This horizontal sweetruler **may** photocopy with less size distortion than the vertical one in Worksheet 2.

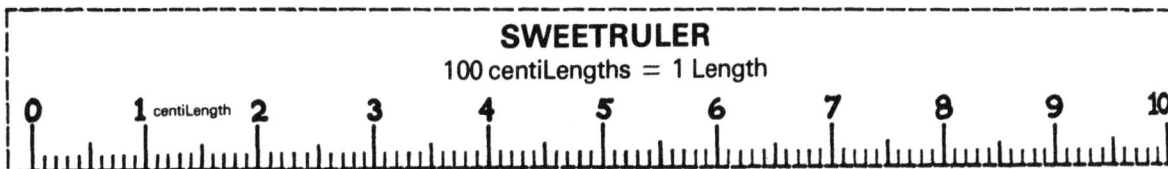

FOR EMERGENCY USE ONLY!

SWEETRULER
100 centiLengths = 1 Length

0   1 centiLength   2   3   4   5   6   7   8   9   10

SUGAR CUBE GRAPH

1 First center your balance. Then weigh these sugar cubes and fill in this table.

VOLUME	MASS
0 sugar cubes	**0 g**
1 sugar cube	
2 sugar cubes	
3 sugar cubes	
4 sugar cubes	
5 sugar cubes	

2 Plot and circle each point. Draw the best possible straight line through your points using a folded index card.

DON'T DRAW INSIDE YOUR CIRCLES

A folded card makes a good straight edge.

MASS (1 sq. = 1 gram)

VOLUME (5 squares = 1 sugar cube)

3 Use an index card to extend your graph line all the way out. Read the mass of 10 sugar cubes.

Go UP, then OVER

MASS

VOLUME

Mass of 10 sugar cubes?

4 Check your prediction using math.

Mass of 10 sugar cubes?

Hint: you know the mass of 5 sugar cubes.

5 Check your prediction on your balance.

10 CUBES   ? GRAMS

Mass of 10 sugar cubes?

6 A box of sugar holds 252 sugar cubes. Find its weight in grams.

TOPS LEARNING SYSTEMS

LIQUID MEASURE

1 Measure a 20 ml volume of water on your balance.

HOW HEAVY IS 20 ml ?

How can you measure *volume* on a balance?

2 Pour this 20 ml of water into another styrofoam cup. Mark *inside* the cup to show how high the water reaches.

20 ml:

MAKE A **SHORT** PENCIL MARK.

POUR INTO ANOTHER CUP

MARK WATER LEVEL

3 Add more water, 20 ml at a time, until you reach 100 ml.

MARK THE NEW WATER LEVEL EACH TIME

— 100 ml
— 80
— 60
— 40
— 20

4 Use your metric cup to find how many milliliters fill a pop can.

Trial 1:   Trial 2:   Trial 3:

How close did you come to the actual volume printed on the can?

FIZZU POP

5 Find the number of ml in a quart milk carton. Stop adding water at the top fold.

MILK

STOP HERE

ONE QUART

LIQUID MEASURE

6 Find the volume of this same milk carton in cubic centimeters.

MEASURE IT!

MILK

DRY MEASURE

ONE QUART

One **ml** of water measures one **cu cm**.
So these two volumes should be about the same.

Find the difference.

Why don't your
LIQUID MEASURE
and
DRY MEASURE
quite agree?

SAVE YOUR MILLILITER MEASURING CUP!

TOPS LEARNING SYSTEMS

GRANULATED DENSITY

1 To complete this table . . .

Mass (g)	0 g	18.2 g	36.4 g
Volume (ml)			

. . . First weigh each mass of *granulated* sugar . . .

18.2g

. . . then pour it into your milliliter cup to find its volume.

2 Plot your points. Use a straight edge (index card) to extend your graph line as far as possible.

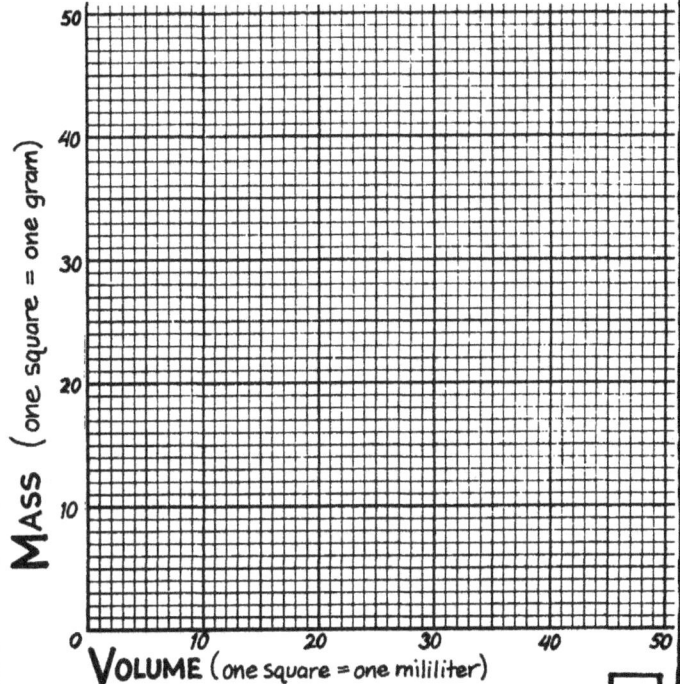

MASS (one square = one gram)

VOLUME (one square = one mililiter)

Teacher Check ☐

3 Read your graph to fill in this column.

GO UP, THEN OVER:

4 Divide each volume into each mass.

Write g/ml (grams in a milliliter) after each number.

V VOLUME (in ml)	M MASS (in g)	$\frac{M}{V}$ OR V)‾M‾ DENSITY (in g/ml)
10 ml		
20 ml		
30 ml		
40 ml		
50 ml		

5 Is a truckful of sugar more dense than a spoonful? Explain.

TOPS LEARNING SYSTEMS

METRIC STRAW

1 How many milliliter cubes of water do you think will fill a straw? Use your metric water cube to make an educated guess.

? YOUR GUESS:

2 Practice pouring straws full of water into a cup. Make sure the straw is completely full and that no water drips into the cup from your hands.

Hold under water at a slant so *all* the air escapes.

Cover both ends with your fingers *before* you lift it from the water.

Allow outside water to drip off.

Take your fingers off *both* ends to empty your strawful.

Keep practicing until you get good at it!   ☐ TEACHER CHECK

3 Center your balance, then weigh a strawful of water. Repeat at least 2 more times to see if you get close to the same value each time.

TRIAL:

1 one strawful = ... grams

2

3

CENTER your balance EACH TIME before you weigh.

Decide what value to use as your

Official Mass:

....

HOW MANY GRAMS?

4 How many *milliliter* of water does your straw hold?

Official Volume:

....

How close was your guess in step 1?

Remember: 1 ml water = 1 g

METRIC WIZ

METRIC WATER CUBE.

SAVE YOUR SODA STRAW ➡

TOPS LEARNING SYSTEMS

THE LAST STRAW

1 Wrap tape around a clean dry straw so it hangs over the end. Squash the tape flat to seal the end.

OVERLAP TAPE

SQUASH FLAT

2 Fill the straw *to the top* with salt.

Dig it into some salt...

... then tap it down.

CLOSED END

TAP TAP

3 Weigh the salt on your centered balance.

Be sure to shake out all the salt.

4 Fill in this table for salt. (You already found the straw's volume.)

Number of Strawsful	VOLUME (in ml)	MASS (in g)	$\frac{M}{V}$ or $V\overline{)M}$ DENSITY (in g/ml)
1			
2			
3			
4			

Decide what value to use as your . . . *Official Density*

5 Graph your results.

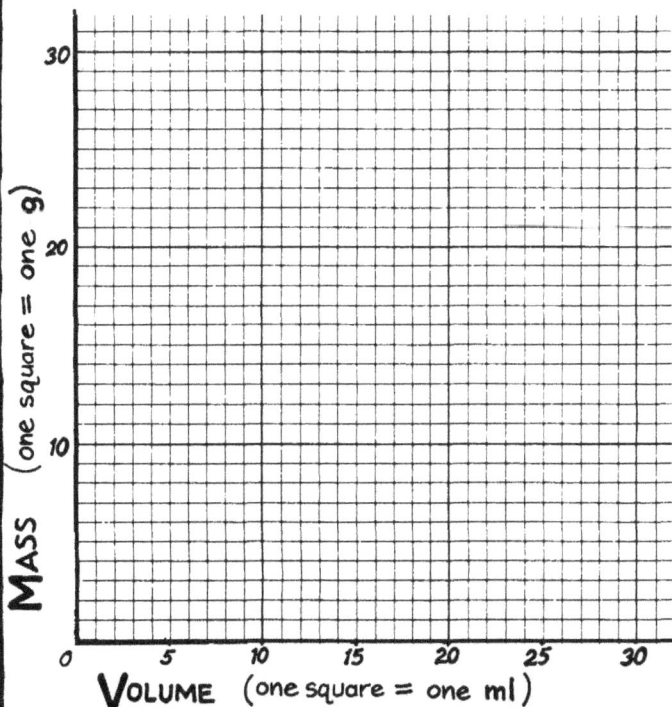

MASS (one square = one g)

30

20

10

0   5   10   15   20   25   30

VOLUME (one square = one ml)

6 Find the density for corn meal. Plot your result on this same graph and label each line.

For your data:

Official Density

TOPS LEARNING SYSTEMS

SINK OR SWIM?

1 You already know these densities. Fill them in:

WATER:	SUGAR:	SALT:	CORN MEAL:
_____ g/ml	_____ g/ml	_____ g/ml	_____ g/ml

2 Complete each little table.

Use your head, not your balance

VOLUME	MASS	VOLUME	MASS	VOLUME	MASS	VOLUME	MASS
0 ml		0 ml		0 ml		0 ml	
20 ml		20 ml		20 ml		20 ml	

3 Draw and label a graph line for each item.

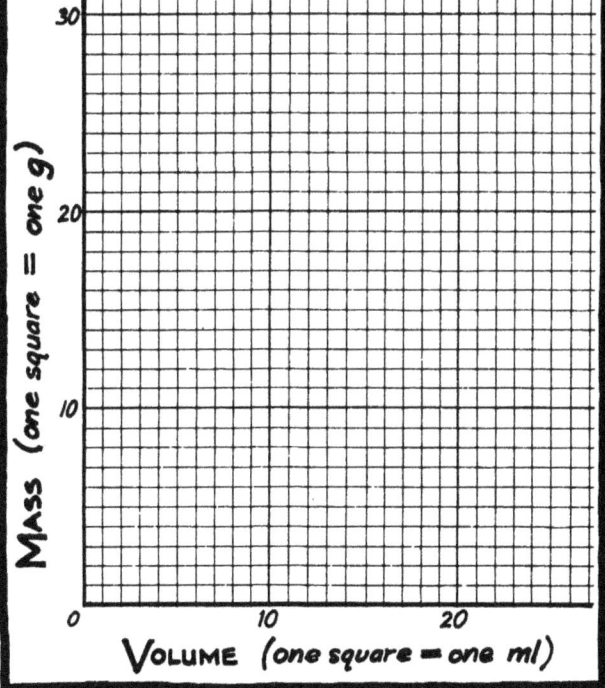

MASS (one square = one g)

30

20

10

0   10   20

VOLUME (one square = one ml)

4 Every material has its own density. What does the slope (steepness of the line) tell you about density?

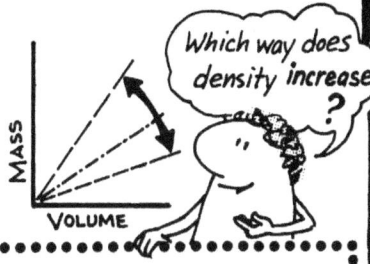

Which way does density increase?

MASS

VOLUME

5 Notice that the density line for water separates salt on one side from sugar and corn meal on the other. Predict what floats in water and what sinks.

6 Test your prediction for sugar. Wrap a cube completely in tape so no corner is left uncovered.

If you see air bubbles come out in water, you have a leak!

Does it float or sink as you predicted?

TOPS LEARNING SYSTEMS

Feedback

If you enjoyed teaching TOPS please tell us so. Your praise motivates us to work hard. If you found an error or can suggest ways to improve this module, we need to hear about that too. Your criticism will help us improve our next new edition. Would you like information about our other publications? Ask us to send you our latest catalog free of charge.

For whatever reason, we'd love to hear from you. We include this self-mailer for your convenience.

Sincerely,

Ron and Peg Marson
author and illustrator

Your Message Here:

Module Title _____ Date _____

Name _____ School _____

Address _____

City _____ State _____ Zip _____

———————————————— FIRST FOLD ————————————————

———————————————— SECOND FOLD ————————————————

RETURN ADDRESS

_____

_____

_____

TOPS Learning Systems
342 S Plumas St
Willows, CA 95988

TAPE HERE

www.ingramcontent.com/pod-product-compliance
Lightning Source LLC
Chambersburg PA
CBHW081511200326
41518CB00015B/2467